# Progress in Probability

Volume 19

# Spatial Stochastic Processes

## *A Festschrift in Honor of Ted Harris on his Seventieth Birthday*

Kenneth S. Alexander
Joseph C. Watkins
Editors

Springer Science+Business Media, LLC

Kenneth S. Alexander
Department of Mathematics
University of Southern California
Los Angeles, CA 90089-1113

Joseph C. Watkins
Department of Mathematics
University of Southern California
Los Angeles, CA 90089-1113

Library of Congress Cataloguing-in-Publication:Data

Spatial stochastic processes : a festschrift in honor of Ted Harris on his seventieth birthday / Kenneth S. Alexander, Joseph C. Watkins, editors.
p. cm. -- (Progress in probability : 19)
Includes bibliographical references.
ISBN 978-1-4612-6766-9 ISBN 978-1-4612-0451-0 (eBook)
DOI 10.1007/978-1-4612-0451-0
1. Stochastic processes. 2. Spatial analysis (Statistics)
3. Harris, Theodore Edward, 1919- . I. Harris, Theodore Edwards, 1919- . II. Alexander, Kenneth S. III. Watkins, Joseph C.
IV. Series.
QA274.S65 1990 89-18290
519.2--dc20 CIP

Printed on acid-free paper.


Originally published by Birkhäuser Boston in 1991
Softcover reprint of the hardcover 1st edition 1991


ISBN 978-1-4612-6766-9

Typeset by Authors in TeX.
9 8 7 6 5 4 3 2 1

# Table of Contents

# Preface

This volume has been created in honor of the seventieth birthday of Ted Harris, which was celebrated on January 11th, 1989. The papers represent the wide range of subfields of probability theory in which Ted has made profound and fundamental contributions.

This breadth in Ted's research complicates the task of putting together in his honor a book with a unified theme. One common thread noted was the spatial, or geometric, aspect of the phenomena Ted investigated. This volume has been organized around that theme, with papers covering four major subject areas of Ted's research: branching processes, percolation, interacting particle systems, and stochastic flows. These four topics do not exhaust his research interests; his major work on Markov chains is commemorated in the standard technology "Harris chain" and "Harris recurrent".

The editors would like to take this opportunity to thank the speakers at the symposium and the contributors to this volume. Their enthusiastic support is a tribute to Ted Harris. We would like to express our appreciation to Annette Mosley for her efforts in typing the manuscripts and to Arthur Ogawa for typesetting the volume. Finally, we gratefully acknowledge the National Science Foundation and the University of Southern California for their financial support.

Kenneth S. Alexander
Joseph C. Watkins

# Introductory Remarks

Murray Rosenblatt

It's a pleasure to be able to give these introductory remarks on Ted Harris and his background. In the process of getting information I was able to get to know Ted and his wife Connie better. Ted Harris was born January 11, 1919, in Philadelphia, Pennsylvania. His parents were Texans and he grew up in Dallas and thought of himself as a Texan. Ted attended the Dallas public school system and spent his first two years in college at Southern Methodist University. The last two years of college were at the University of Texas, Austin, and he received a Bachelor of Arts degree from that University. It is interesting and amusing to reflect, Ted's initial two and one-half years of graduate work were with R. L. Moore at the University of Texas, Austin, and were in the direction of point set topology.

World War II then intervened and Ted entered the Army Air Force in 1942 as a weather officer. He spent two and one-half years in England (1943–45), including some time at High Wycombe, at the 8th Air Force Headquarters. While there he met Ted (J. W. T.) Youngs and Edwin Hewitt. After the war, Ted thought of meteorology as a profession and consulted someone at the Weather Bureau. The suggestion was a degree in statistics. He went to Princeton in the fall of 1945 to work with Sam Wilks. It is curious that Wilks suggested he work on branching processes. At that time there were a number of contemporary graduate students who have since become quite well-known. Among these are Richard Bellman, Kai Lai Chung, Steve Diliberto, Leon Henkin, Gilbert Hunt, Samuel Karlin, Paul Meier, and Harold N. Shapiro. It is worth noting that Harold Cramér of the University of Stockholm visited Princeton University in the academic year 1946–47 and lectured there. Ted Harris received his Ph.D. degree in 1947. 1947 was also especially noteworthy since Ted Harris and Connie Feder were married in that year. They've been happily married for 43 years and have two children, Marcia and Stephen.

Ted and Connie came out to the West Coast in 1947 when Ted took a position at the Rand Corporation. At that time John Williams was head of the mathematics group that Ted had joined. There were many consultants visiting Rand but among them were people like Ken Arrow, David Blackwell, Samuel Karlin, Jimmie Savage, Tracy Thomas, John Von Neumann and J. W. T. Youngs. Richard Bellman, George Dantzig and Lloyd Shapley were on the staff. Ted visited Columbia University in the Department of Statistics in 1953 as Associate Professor and collaborated with Herbert Robbins on ergodic theory for Markov chains. In 1963 he visited Stanford University as Professor in the Department of Statistics. From 1959 to 1965 Ted was head of the mathematics group at Rand. In 1966 he moved to the University of Southern California as Professor of Mathematics and Electrical Engineering. He had six Ph.D. students in the first ten years and has three doctoral students now.

In the period 1955 to 1958 Ted Harris served as editor of the Annals of Mathematical Statistics, and he was elected as president of the Institute of Mathematical Statistics for the year 1966–67. He started the Southern California Probability Symposium and this 19th meeting held in honor of his 70th birthday is a proper acknowledgement of his dedication and of his research.

Ted Harris has contributed with great originality, power, and depth to the theory and application of stochastic processes. In his early research, he concentrated on a number of concrete classes of processes. In some cases he determined and generally characterized first passage and recurrence time distributions for a large class of random walks. Later he extended work of Wolfgang Doeblin (a son of the author Alfred Doeblin) in a fundamental way and obtained necessary and sufficient conditions for the existence of a $\sigma$-finite invariant measure for a large class of recurrent Markov processes. The Harris invariant measure is basic in the use of renewal type analyses in the case of a general state space. Harris' monograph, "The Theory of Branching Processes" [1963], was a landmark and is still a basic book for the field. This book incorporated much of Harris' own work in the area and was also a synthesis of related research. Branching processes serve as models for population growth, neutron interaction, and electron-photon cascades. His 1960 paper on percolation was an early seminal contribution. It contained most of the ideas for rigorous determination of some critical probabilities and stimulated much research in the field. Recently he developed an extensive theory concerned with a class of interacting particle systems including additive set-valued Markov processes, symmetric simple exclusion models and certain contact models. His most recent work deals with stochastic flows involving applications to models based on stochastic differential equations. Ted contributed important new ideas at an early point of development of each of the areas he worked in. His important contribution was acknowledged in his recent election to the National Academy of Sciences, and in his honorary doctorate from Chalmers Institute of Technology in Sweden.

He is known to many of us as a sensitive and congenial individual and we have participated in this meeting as a sign of our warm appreciation of him.

University of California, San Diego

# Publication List of T. E. Harris


1. *Note on differentiation under the expectation sign*, Ann. Math. Statist. **18** (1947), pp. 294–295.
2. *Branching Processes*, Ann. Math. Statist. **19** (1948), pp. 474–494.
3. *Age-dependant stochastic branching processes*, with R.Bellman, Proc. Nat. Acad. Sci. **34** (1948), pp. 601–604.
4. *Timing of events between observations*, with P. Meier and J. W. Tukey, Human Biology **22** (1950), pp. 249–270.
5. *Recurrence time for the Ehrenfest model*, with R. Bellman, Pacific J. Math. **1** (1951), pp. 179–193.
6. *The Monte Carlo Method: Estimation of Particle Transmission*, with H. Kahn, NBS Applied Math. Series **12** (1951), pp. 27–30.
7. *Optimal Inventory policy*, with K. Arrow and J. Marschak, Econometrica **19** (1951), pp. 250–272.
8. *Some mathematical models for branching processes*, in "Proc. Second Berkeley Symposium," 1951, pp. 305–328.
9. *First passage and recurrence distributions*, Trans. Amer. Math. Soc. **73** (1952), pp. 471–486.
10. *On age-dependent binary branching processes*, with R. Bellman — expansion of results announced in [3] above, Ann. of Math. **55** (1952), pp. 280–295.
11. *Ergodic theory of Markov chains*, with H. Robbins, Proc. Nat. Acad. Sci. **39** (1953), pp. 860–864.
12. *On chains of infinite order*, Pacific J. Math 5, Suppl. **1** (1955), pp. 707–724.
13. *The existence of stationary measures*, in "Proc. Third Berkeley Symposium," 1956, pp. 113–124 of Vol. 2..
14. *The random functions of cosmic ray casades*, Proc. Nat. Acad. Sci. **43** (1957), pp. 509–512.
15. *Transient Markov chains with stationary measures*, Proc. Amer. Math. Soc. **8** (1957 937–942).
16. *A mathematical model for multiplication by binary fission*, in "The Kinetics of Cellular Proliferation," F. Stohlman, Edl, Grune and Stratton, 1959, pp. 368–381.

17. *A lower bound for the critical probability in a certain percolation process*, Proc. Cambridge Philos. Soc. **56** (1960), pp. 13–20.
18. "The Theory of Branching Processes," Volume 119 of the series Grundlehren der Math. Wisc., Springer, 1963. Published in U. S. by Prentice-Hall.
19. *Diffusion with collisions between particles*, J. Appl. Probability **2** (1965), pp. 323–338.
20. *Infinite Product Markov processes*, TAMS **130** (1968), pp. 141–152.
21. *Counting measures, monotone random set function, Z*, Wahrscheinlichketstheorie und Verw. Gebiete **10** (1968), pp. 102–119.
22. *Random measures and motions of point processes. Z.*, Wahrscheinlichketstheorie und Verw. Gebiets **18** (1971), pp. 85–115.
23. *Nearest neighbor interaction processes on multi-dimensional lattices*, Advances in Math. **9** (1972), pp. 66–89.
24. *Contact interactions on a lattice*, Ann. Prob. **2** (1974), pp. 969–988.
25. *On a class of set-valued Markov processes*, Ann. Prob. **4** (1976), pp. 175–194.
26. *A correlation inequality for Markov processes in a partially ordered space*, Ann. Prob. **5** (1977), pp. 451–454.
27. *Additive set-valued Markov processes and graphical methods*, Ann. Prob. **6** (1978), pp. 355–378.
28. *Brownian motions on the homeomorphisms of the plane*, Ann. Prob. **9** (1981), pp. 232–254.
29. *Coalescing and non-coalescing stochastic flows in R1*, Stochastic Processes and Applications **17** (1984), pp. 187–210.
30. *Isotropic stochastic flows and a related property of non-random potential flows*, in "Proceedings of the Fifteenth Conference on Stochastic Processes and Applications, Nagoya, 1985," Lecture Notes in Mathematics, Springer Verlag, 1986.
31. *Isotropic stochastic flows*, joint with Peter Baxendale, Annals of Probability **14** (1986), pp. 1155–1179.

# Branching Processes

# Branching Random Walk

Peter Ney

## 1. Introduction

In a volume dedicated to Ted Harris, it is appropriate that there should be some discussion of branching processes, a subject of which he is one of the founders. In a series of papers in the 1940's and 50's (see references [1] to [9] at the end of this paper), culminating in his famous 1963 book "The Theory of Branching Processes" [10], he helped to lay the rigorous mathematical foundations of the subject, to answer a number of basic questions, and to show the direction of many future lines of research.

In this note I will give a short survey of one line of development, namely branching random walks. These models had their origins in the theory of cascade processes, which (naturally) were first studied by physicists (in the 1930's; see e.g. [16], [13]). The last chapter of Harris' book is a rigorous treatment of some aspects of this subject which provided an important stimulus for subsequent mathematical work.

In section 2, I will make a few remarks on these cascades, and on how they relate to branching random walks. Section 3 contains a short discussion of the Central Limit Problem for branching random walks. Section 4 summarizes some work on questions of large deviation type. In section 5, I give the outline of a new result on multi-type branching random walks, which describes how the composition of the population varies throughout the space in which the particles move. This is joint work with M. Bramson and J. Tao. The details of this work are contained in our paper [23]. Section 6 refers to a question on the electron-photon cascade mentioned in Harris' book.

## 2. Cascades

Among the models of cascade processes that have been extensively studied are the electron-photon cascade and the neucleon cascade. In the former electrons give rise to photons as they pass through a medium, this process being called "bremsstrahlung". The energy of the electron is reduced by the

1700 Main Street
Santa Monica, California

October 1, 1957

Mr. Peter E. Ney
468 Riverside Drive
New York 27, New York

Dear Mr. Ney:

It would take some time to get my draft chapter on nuclear cascades ready for release by RAND. Accordingly I am sending you an outline of the material. I don't think there is much new in it, although a few points may suggest new lines of work.

— — — . — . — —
— — — — — —

p 3

b. Behavior for $x$ near $2\lambda t\mu$. Suppose we write the series for $m^*(x,t)$ as $m^*(x,t) = e^{-x} m(e^{-x}, t)$

$$(9) \quad \cdots \quad e^{\lambda t} \sum_{r=1}^{\infty} p_r k_r^*(x)$$

where ~~pr~~ $p_r$ is the probability that a Poisson – distributed random variable with mean $2\lambda t$ takes the value $r$. The Poisson probabilities are maximum near $r = 2\lambda t$. Now when $r$ is large, $k_r^*$ has approximately the Gaussian form with mean $r\mu$ and variance $r\sigma^2$, provided $x$ is near $r\mu$. Using the form of the central limit theorem on p. 166 (84) of Khinchin's Statistical Mechanics we get

$$m^*(x,t) \sim \frac{e^{\lambda t}}{\sqrt{4\pi\lambda t(\sigma^2+\mu^2)}} \exp\left\{-\frac{1}{2}\frac{(x-2\lambda t\mu)^2}{2\lambda t(\sigma^2+\mu^2)}\right\}$$

Figure 1

Figure 1

-4-

as $t \to \infty$ provided $(x - 2\lambda t\mu)t^{-\frac{1}{2}}$ remains bounded. Thus for large $t$ we expect half the total particles to have energies greater than $e^{-2\lambda t\mu}$.

Conjecture. Suppose $(x - 2\lambda t\mu)t^{-\frac{1}{2}} = (-\log E - 2\lambda t\mu)t^{-\frac{1}{2}}$ remains bounded as $t \to \infty$. Then $N(E,t) / N(0,t)$ converges in probability to

$$(10) \quad \Phi\left(\frac{-\log E - 2\lambda t\mu}{\sqrt{2\lambda t(\sigma^2 + \mu^2)}}\right)$$

where $\Phi$ is the cumulative Gaussian.

I think this would make a nice thesis result. Perhaps you could proceed by finding the ~~covariance~~ correlation between $N(E_1,t)$ and $N(E_2,t)$.* Note that $N(0,t) / e^{\lambda t}$ converges to a random variable.**

It is also of some interest to study the total energy $Z(t)$ in the cascade. If $\phi(s,t)$ is the Laplace transform of the total energy at $t$, then

$$(11) \quad \frac{\partial\phi}{\partial t} = -\lambda\phi + \lambda\iint \phi(su,t)\,\phi(sv,t)\,k(u,v)\,du\,dv.$$

If $b = 1 - 2\int_0^1 uk(u)du$, notice that $e^{bt} Z(t)$ $= W(t)$ is a martingale, and converges with probability 1 to a random variable $W$. You might be interested in studying the distribution of $W$. If the second moment of $W(t)$ is bounded as $t \to \infty$ (you can find conditions on $k$ that imply this) then $W$ is nondegenerate — i. e., not identically 0. If the 2nd moment isn't bounded, I don't know.

I have a little material on the generating function for $N(E,t)$. As you know it satisfies

* Presumably it approaches 1 provided $E_1$ and $E_2$ are close to $e^{-2\lambda t\mu}$

** From an elementary result on branching processes

Figure 1, cont.

acquired energy of the photon. Photons in turn give rise to two electrons, resulting in more photon production, etc. The object of interest is the number of electrons and photons in various energy ranges, namely $(N_1(E,t), N_2(E,t)) =$ the number of electrons and photons, respectively, of energy $\geq E$ at depth or time $t$.

This can be thought of as a binary two type branching process, with a type 1 parent producing a type 1 and a type 2 offspring and a type 2 parent producing two type 1's. The "lifetimes" are exponentially distributed. A potential application of our result in section 5 is to such processes. For careful definitions and construction see chapter VII of Harris [10].

The nucleon cascade is a single type process. A parent of energy $E$ gives rise to two offspring of energies $E_1, E_2$, with joint distribution $\varphi(E_1, E_2 \mid E_0)$, after living a random time that is exponentially distributed with parameter $\lambda$. Assume that $\varphi$ is symmetric in $(E_1, E_2)$ and is a homogeneous function in the sense that $\varphi(cE_1, cE_2 \mid cE_0) = \varphi(E_1, E_2 \mid E_0)$ for all $c > 0$; and write $\varphi(E_1, E_2 \mid 1) = \varphi(E_1, E_2)$. Assume from now on that the original parent particle has energy 1. Let $N(E,t) =$ the number of particles of energy at least $E$ at time $t$. Then by a standard decomposition of the state space according to the possible outcomes at the first time of particle production, the generating function $Q(z,t,E) = Ez^{N(E,t)}$ satisfies the equation

$$(2.1) \quad \frac{1}{\lambda}\frac{\partial Q(E,tz)}{\partial t} + Q(E,t,z) = \int_E^1 \int_E^1 \varphi(du_1, du_2) Q\left(\frac{E}{u_1}, t, z\right) Q\left(\frac{E}{u_2}, t, z\right).$$

This is called the Janossy $G$-equation (see [27]). The generating functions of $N_i(E,t)$ $i = 1,2$ for the electron-photon cascade satisfy a similar pair of coupled integro-differential equations (see Harris [10]). Much of the early study of cascades was focused on such equations, including existence, uniqueness, approximation of solutions, properties of moments, series expansions, etc. ([32], [33], [41], [36], [37]). In these studies $E$ was usually held fixed, and thus $P(N(E,t) = 0) \longrightarrow 1$ as $t \to \infty$. The rates of convergence of such quantities were studied, as well as of ratios like $P(N(E,t \mid E_0) = i)/P(N(E,t \mid E_0) = j)$.

In 1957, T. Harris proposed a class of problems in which $E = E_t$ depended on $t$ and $E_t \to 0$ at a suitable rate as $t \to \infty$. In a private communication to this author (while I was a graduate student at Columbia), he pointed out that the mean $m(E,t) = EN(E,t)$ satisfies

$$e^{-x_t} m(e^{-x_t}, t) \sim \frac{e^{\lambda t}}{\sqrt{4\pi\lambda t(\sigma^2+\mu^2)}} \exp\left\{-\frac{1}{2}\frac{(x_t - 2\lambda t\mu)^2}{2\lambda t(\sigma^2+\mu^2)}\right\} \tag{2.2}$$

as $t \to \infty$, where $x_t$ is a function of $t$ such that $(x_t - 2\lambda t\mu)t^{-1/2}$ remains bounded, and $\mu$ and $\sigma^2$ are the mean and variance of $-\log E_1$.

Professor Harris proposed that I study the limit theory of $N(x_t,t)/N(t)$ where $N(t)$ = the total number of particles at time $t$. In Figure 1, I display parts of his 1957 letter to me, and I take this opportunity to thank him for his help during my early struggles in this subject.

The result was (see [38]) (as he conjectured) that if $\mu$ and $\sigma^2$ exist, and $x_t = 2\lambda\mu t - \gamma[2\lambda t(\mu^2+\sigma^2)]^{1/2}$, then

$$\frac{N(x_t,t)}{N(t)} \longrightarrow \text{the Gaussian} \quad \Phi(\gamma) \tag{2.3}$$

in probability.

In the interest of historical accuracy one should point out that as with so much else in probability theory A. N. Kolmogorov [30] was the first to study such models. He proved a discrete time version of (2.2), and stated (without proof) the discrete form of a slightly more special version of (2.3).

## 3. Branching random walk

A simple logarithmic transformation, replacing $(E_1, E_2)$ by $(-\log E_1, -\log E_2)$, converts the cascade process into what we now call a branching random walk. The multiplicative homogeneity of the energy distribution is now replaced by the additive or "spatial" homogeneity of the random walk. At little cost one can introduce some extra generality: namely from binary fission to general particle production with $p_k$ = probability of $k$ offspring, $k = 0, 1, \ldots$. Also for many purposes a general lifetime distribution $G(\cdot)$ can be allowed (the so-called age-dependent case), and the spatial movement can be considered in $d \geq 1$ dimensional space.

Let $Z(\Gamma, t)$ = the number of particles in $\Gamma \subset \mathbb{R}^d$ at time $t$. When $\Gamma$ is the orthant $\{y \in \mathbb{R}^d : y_i \leq x_i,\ i = 1, \ldots, d, x \in \mathbb{R}^d\}$ we write $Z(x,t)$ for short. Also write $Z(\mathbb{R}^d, t) = Z(t)$. Assuming that $m = \Sigma k p_k > 1$ (supercriticality) and letting $\alpha$ be the Malthusian parameter determined by $m \int e^{-\alpha t} G(dt) = 1$, one can show under second moment hypothesis on the spatial motion that

$$e^{-\alpha t}[Z(x_t, t) - Z(t) \cdot \Phi(\gamma)] \longrightarrow 0 \tag{3.1}$$

in mean square, where $\vec{x}_t = \vec{\mu}t + c\vec{\gamma}\sqrt{t}$, $\vec{\mu} \in \mathbb{R}^d$ is the mean of the marginal d.f. of the spatial motion, $c =$ a constant depending on the first two moments of the given spatial d.f.'s, $\vec{\gamma} \in \mathbb{R}^d$, and $\Phi(\cdot)$ is a $d$-dimensional Gaussian d.f. with zero mean. We proved this fact (in [39]) under a second moment condition on the particle production $\{p_k\}$.

As a corollary one also gets that

$$\frac{Z(\vec{x}_t, t)}{EZ(t)} \overset{\text{m.s.}}{\longrightarrow} \vec{W}\Phi(\gamma), \tag{3.2}$$

where

$$W = \lim \frac{Z(t)}{EZ(t)} \tag{3.3}$$

is a familiar random variable in ordinary age-dependent branching theory.

Of course there are analogous results in discrete time, where $G(\cdot)$ puts unit mass at 1.

A necessary and sufficient condition for many of the basic limit theorems for supercritical branching processes is the existence of the logarithmic moment $\Sigma(k \log k)p_k$, of the particle production. Thus a fair amount of effort went into weakening the hypothesis for (3.2) to the same moment condition. Asmussen and Kaplan proved that the discrete time version of (3.2) (with $d = 1$) holds for convergence in probability under $\Sigma(k \log k)p_k < \infty$, and for probability one convergence under $\Sigma k(\log k)^{1+\epsilon} p_k < \infty$ with $\epsilon > 0$ [12]. Finally Athreya and Kaplan proved the desired result with prob. 1 convergence under $\Sigma(k \log k)p_k < \infty$ [13].

An alternative approach to the above is via martingales. For the ordinary Galton-Watson process with $Z_n =$ number of particles at time $n$ and $m = EZ_1$, $W_n = Z_n/m^n$ is a martingale and converges $\Sigma(k \log k)p_k < \infty$. This suggests the existence of martingales for other branching models, including the branching random walk (see e.g. [14], [29] or [18] ). Let us describe this martingale in the discrete time case. Write $Z^n(\Gamma)$ for the number of $n$th generation particles in the Borel set $\Gamma \subset \mathbb{R}^1$, with $Z^0(\Gamma) = \delta_0(\Gamma)$ (i.e. start the process at the origin with a single particle). Let

$$m(\Gamma) = EZ^1(\Gamma) \quad \text{and assume} \quad m = m(\mathbb{R}^1) > 1.$$

Define the generating function

$$\lambda(\alpha) = \int e^{\alpha s} m(ds) \tag{3.4.1}$$

and let $\Lambda(\alpha) = \log \lambda(\alpha)$. Also write $\frac{\lambda(\alpha)}{m} = \varphi(\alpha)$ for the generating function of the mean motion of a single particle, and $\mu = \varphi'(0)$. Now

$$0 \le W_n(\alpha) = e^{-n\Lambda(\alpha)} \int e^{\alpha t} Z^n(dt) \tag{3.4}$$

is a martingale ([14], [18]) and thus

$$\lim_{n\to\infty} W_n(\alpha) = W(\alpha) \quad \text{exists a.s. and is} \quad \le 1 \quad \text{(by Fatou).} \tag{3.5}$$

Assume

$$O \in \text{In } \mathcal{D} = \{\alpha : \Lambda(\alpha) < \infty\}.$$

Biggins [18] has proved that for $\alpha \in$ In $\mathcal{D}$

$$EW(\alpha) = 1 \quad \text{if and only if} \quad EW_1(\alpha)|\log W_1(\alpha)| < \infty. \tag{3.6}$$

The following idea is due to A. Joffe (private communication). Suppose that one can prove that the convergence in (3.5) is uniform for $\alpha$ in a neighborhood of 0. Then

$$W_n\left(\frac{\alpha}{\sqrt{n}}\right) \xrightarrow{\text{a.s.}} W(0) = W \qquad \text{(in (3.3)).} \tag{3.7}$$

But

$$W_n\left(\frac{\alpha}{\sqrt{n}}\right) = \frac{\int e^{\frac{\alpha}{\sqrt{n}}s} Z_n(ds)}{m^n \varphi^n\left(\frac{\alpha}{\sqrt{n}}\right)}. \tag{3.8}$$

Assuming that the measure associated with the average spatial motion has zero mean and unit variance, $\varphi^n\left(\frac{\alpha}{\sqrt{n}}\right) \longrightarrow \exp\{-\alpha^2/2\}$.

Reducing the discussion that led to (3.1)–(3.3) to discrete time ($G(\cdot)$ puts unit mass at 1) with $\mu = 0$, the numerator in (3.8) is just the generating function of $Z^n(x\sqrt{n})$ and $EZ(n) = EZ^n(\mathbf{R}^1) = m^n$. Thus the generating function of the LHS of (3.2) converges to $W$ times the generating function of $\Phi(\cdot)$ w.p.1, in a neighborhood of 0, and (3.2) follows. Joffe has carried out this argument under a second moment condition on $\{p_k\}$ by considering $W_n(\cdot)$ as a vector valued martingale on the space of continuous functions in an interval about 0, using techniques similar to those in Joffe, Le Cam, Neveu [28]. It will appear in the Proceedings of the 1989 Vilnius Conference.

There are a number of other models that yield results similar to those above. Among them are the so-called branching diffusions where particles multiply as before but move in accordance with a diffusion. See e.g. Asmussen and Herring [11] for a proper definition of these processes. An interesting result of S. Watanabe [43] is the following. Let $\Gamma$ be a bounded domain $\subset \mathbf{R}^d$ and $|\Gamma| =$ the Lebesgue measure of $\Gamma$. Assume $p_0 = 0$, $\Sigma k p_k > 1$ and $\Sigma k^2 p_k < \infty$. Let $Z_t(\Gamma) =$ the number of particles in $\Gamma$ at time $t$. Using a martingale argument Watanabe proved the "local" limit theorem.

$$(3.9) \qquad (2\pi t)^{d/2}\frac{Z_t(\Gamma)}{Z_t(\mathbf{R}^d)} \longrightarrow |\Gamma| \qquad \text{w.p.1.}$$

## 4. Behavior away from the mean

The results of the previous section describe the behavior of $Z^n(\Gamma_n)$ where $\Gamma_n$ is a set of the form $\Gamma_n = \vec{\mu}n + \sqrt{n}\Gamma$ for some fixed $\Gamma \subset \mathbf{R}^d$ or (in the case of a local result like Watanabe's) where $\Gamma_n = \vec{\mu}n + \Gamma$. In either case we were concerned with the number of particles in a region about the central tendency of the process $\vec{\mu}n$. Most of the particles are in $\vec{\mu}n + \sqrt{n}\Gamma$, their expected number being of order $m^n$.

In the case of a supercritical process the population will, however, also grow in regions away from this center of mass, though of course the growth rate will be less than $m^n$. In particular we are interested in the asymptotic behavior of $Z_n(n\Gamma)$ for $\Gamma \subset \mathbf{R}^1$. (We limit ourselves in this exposition to processes on the line.) Recall that $\mu = m^{-1}\int sm(ds)$ is the mean of the motion of a single particle. Now if $\Gamma$ is an open interval and $\mu \in \Gamma$ then $Z^n(n\Gamma) \sim Z^n(\mathbf{R}) = Z^n$. But if $\mu \notin \Gamma$ we will have a different behavior, and this is what we want to

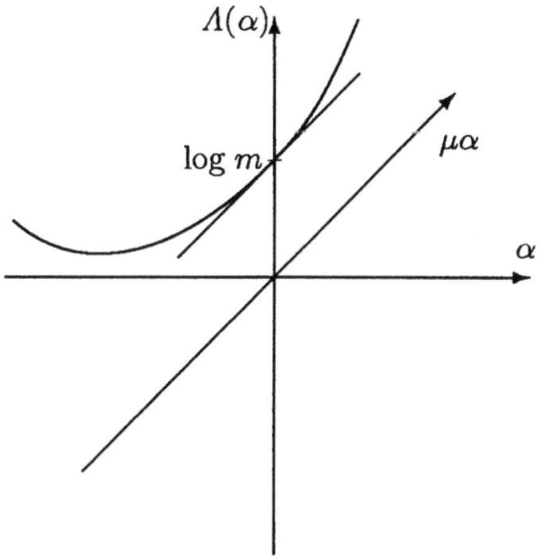


Figure 2a

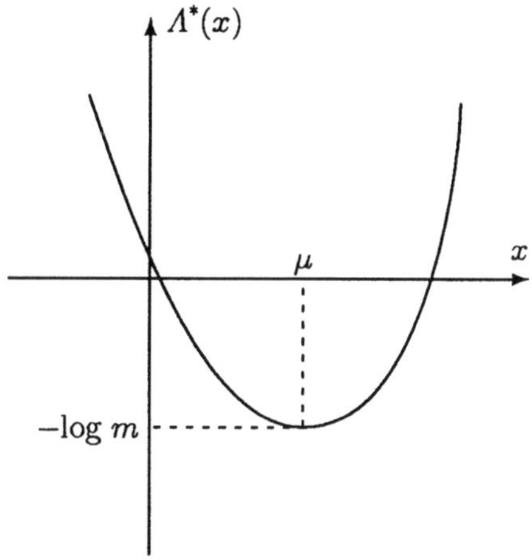


Figure 2b

look at. It is natural that the tools needed in such a study are a combination of large deviation (LD) ideas with branching methodology.

To start with it is easy to check that $EZ^n(\Gamma) = m^{*n}(\Gamma) =$ the $n$-fold convolution of $m(\cdot)$. Since $m(\cdot)$ is a finite measure we have by standard LD theory that if $\lambda(\alpha) < \infty$ in a neighborhood of 0, then

$$(4.1) \qquad \frac{1}{n} \log m^{*n}(n\Gamma) \text{ ``} \longrightarrow \text{''} \; - \inf_{x \in \Gamma} \Lambda^*(x),$$

where by "$\longrightarrow$" I mean in the usual LD sense: lim sup of the LHS is $\leq$ the RHS for $\Gamma$ a closed set and lim inf LHS $\geq$ RHS for $\Gamma$ open. Here $\Lambda^*(x) = \sup[x\alpha - \Lambda(\alpha)]$ is the convex conjugate of $\Lambda(\cdot)$. Note that due to the assumed supercriticality ($m \geq 1$), $\Lambda^*(\mu) < 0$, since $\Lambda^*(\mu) = \sup[\mu\alpha - \Lambda(\alpha)]$ is achieved for $\alpha$ satisfying $\Lambda'(\alpha) = \mu$, i.e. $\alpha = 0$ (see Fig. 2).
Also it is clear from Fig. 2a and the convexity of $\Lambda(\cdot)$ that $\Lambda^*(x) \geq \Lambda^*(\mu)$ for all $x$, hence Fig. 2b. This is important, because $\{x : \Lambda^*(x) < 0\}$ will be the region of applicability of our results.

If $\Gamma$ is an interval then "$\longrightarrow$" in (4.1) can be replaced by a legitimate limit. The following result is the analogous first order LD theorem for the branching process itself.

**Theorem 4.1.** (J. D. Biggins [17]) *Assume that* $\lambda(\alpha) < \infty$ *in a neighborhood of* 0. *Suppose* $a < \mu$. *If* $\Lambda^*(a) < 0$ *then*

$$\frac{1}{n} \log Z^n((-\infty, an]) \longrightarrow -\Lambda^*(a) \qquad a.s.. \tag{4.2}$$

*If* $\Lambda^*(a) > 0$, *then* $Z^n((-\infty, an]) = 0$ *for all but a finite number of* $n$.

In the standard LD theory a variety of sharpenings of the logarithmic limit theory are available, going back to the asymptotic expansion of Bahadur and Rao [15]. Similarly, there are asymptotic versions of (4.2). These are important for us, since the parameters in the result we will give in the next section do not appear at the logarithmic level.

**Theorem 4.2.** (Biggins [18], [20]) *Assume that* $0 \in \mathcal{D} = \{\alpha : \Lambda(\alpha) < \infty\}$. *Let* $W_m(\alpha)$ *be the martingale (3.4) and* $W(\alpha)$ *be its limit.*

(1) For $\alpha \in \mathcal{D}$, $EW(\alpha) = 1$ if and only if $EW_1(\alpha) \log W_1(\alpha) < \infty$.
(2) Assume that the solution $\alpha_a$ of $\Lambda'(\alpha) = a$ exists. If $\Lambda^*(a) < 0$, $EW_1(\alpha_a) |\log W(\alpha_a)|^{\frac{5}{2}+\epsilon} < \infty$ for some $\epsilon > 0$, and $e^{\alpha_a x} g(x)$ is directly Riemann integrable, then

$$\sigma\sqrt{2\pi n} e^{\Lambda^*(a)n} \int g(x - an) m^{*n}(dx) \longrightarrow \int e^{\alpha_a x} g(x) dx. \tag{4.3.1}$$

and

$$\sigma\sqrt{2\pi n} e^{\Lambda^*(a)n} \int g(x - an) Z^n(dx) \xrightarrow{\text{a.s}} W(\alpha_a) \int e^{\alpha_a x} g(x) dx. \tag{4.3.2}$$

where $\sigma^2 = \Lambda''(\alpha_a)$.

This is a "local" limit theorem. For example, if $\Gamma$ is an interval, then it implies (by taking $g = \chi_\Gamma$) that

$$\frac{Z^n(an + \Gamma)}{m^{*n}(an + \Gamma)} \longrightarrow W(\alpha_a) \qquad \text{a.s.} \tag{4.4}$$

A similar result to (4.3) when particle motion is in $\mathbb{R}^d$, $d \geq 1$ was obtained (in the continuous time Markov case) by K. Uchiyama [41]. He proved, under $\Sigma k^3 p_k < \infty$, and standard LD hypotheses on the generating function of particle motion, that

$$t^{d/2} e^{\Lambda^*(a)t} Z^t(at + \Gamma) \longrightarrow cW(\alpha_a) \qquad \text{a.s.,} \tag{4.5}$$

where $c$ is a constant depending on $\underline{a}$ and $\Gamma$.

To prove theorem 4.1 Biggins used Borel-Cantelli in one direction and a "pruning" argument on the branching tree in the other. An alternative argument using the martingale $W_n(\alpha)$ also works nicely (see B. Chauvin [25], J. Neveu [35]).

I just mention briefly a few further interesting results on the large deviation of branching r.w.'s. A lot is known on the "shape" or "range" of the process. In $\mathbb{R}^d$, there are sharp results by M. Bramson [21], who proved for branching Brownian motion that (subject to $\Sigma k^2 p_k < \infty$) the maximal displacement $M(t)$ satisfies

$$P\{M(t) \leq \sqrt{2t} - 3.2^{-3/2} \log t + b_\epsilon(t)\} = \epsilon \tag{4.6}$$

for a function $b_\epsilon(t) = O(t)$ as $t \to \infty$. "Weak law of large numbers" versions of such results for the so-called first birth time of an age-dependent process and for branching random walks, were first proved by Hammersley [26] and Kingman [29]. For a simple branching random walk, Bramson [22] proved (under $\Sigma k^{2+\delta} p_k < \infty$) that $M_n =$ the minimum displacement satisfies

$$M_n - (\log 2)^{-1}[\log \log n - \log V + o(1)] \to 0 \qquad \text{w.p.1,} \tag{4.6a}$$

where $V$ is a random variable.

Biggins [19] also showed that for motion in $\mathbb{R}^d$ the asymptotic shape of the random walk is (in a suitable sense) the set $\{x : \Lambda^*(x) \leq 0\} = \mathcal{I}$. Outside $n\mathcal{I}$ the number of particles $\to 0$ as $n \to \infty$; for open subsets $B$ of $\mathcal{I}$, the number in $nB$ grows exponentially.

Generalizing in a different direction, Laredo and Ronalt [31] studied an age-dependent branching r.w. The results are similar to those in the discrete time case, but a "generalized Malthusian parameter" comes into the construction of the rate function in an interesting way.

## 5. The composition of the population in the multi-type case

The results in this section are joint work with M. Bramson and J. Tao. We consider the multi(d)-type version of the process in section 4. To describe the model we start with a standard multi-type Galton-Watson process as defined, for example, in chapter II of [10], and let a particle of type $j$ undergo a random spatial motion with distribution $G_j(\cdot)$ in its lifetime. All motions are independent r.v.'s. We treat here the case where the motions are lattice (integer) valued 1-dimensional random variables. The object of interest is $Z_{ij}^n(\Gamma)$ = the number of type $j$ particles in the set $\Gamma \subset \mathbb{Z}$ at time $n$ descended from a type $i$ ancestor of time 0. ($\mathbb{Z}$ = the integers.) If $\Gamma = \{x\}$, the single point set, we write $Z_{ij}^n(x)$ to denote the number of particles at the nearest integer to $x$. Let $Z_i^n(\Gamma) = (Z_{i1}^n(\Gamma), \dots, Z_{id}^n(\Gamma))$ = the population vector at $n$, and $Z_i^n(\Gamma) = \sum_j Z_{ij}^n(\Gamma)$.

Our objective is to study the asymptotic behavior of $Z_i^n(an)$, and in particular of the ratios $Z_{ij}^n(an)/Z_{k\ell}^n(an)$, where $\underline{a}$ is a fixed point $\in \mathbb{R}^n$.

**Notation.** We will use the following notation: $m_{ij}(\Gamma) = EZ_{ij}(\Gamma)$, $m_{ij} = m_{ij}(\mathbb{Z})$, $M(\Gamma) =$ the matrix $\{m_{ij}(\Gamma)\}$, $M^n(\Gamma) = \{m_{ij}^n(\Gamma)\} = \{EZ_{ij}^n(\Gamma)\}$, etc. Also

$$
\begin{aligned}
\widehat{m}_{ij}(\alpha) &= \int e^{\alpha x} m_{ij}(dx), \quad \widehat{M}(\alpha) = \{\widehat{m}_{ij}(\alpha)\},\\
\lambda(\alpha) &= \text{ largest eigenvalue of } \widehat{M}(\alpha), \quad \lambda(0) = \rho,\\
\Lambda(\alpha) &= \log \lambda(\alpha);\\
\overline{m}_{ij} &= \int \mathrm{s}m_{ij}(ds), \quad \overline{M} = \{\overline{m}_{ij}\}, \quad \Lambda'(0) = \mu = \rho^{-1}\ell(0)\overline{M}r(0);
\end{aligned}
$$

$\ell(\alpha) = (\ell_1(\alpha), \dots, \ell_d(\alpha))$ and $r(\alpha) = (r_1(\alpha), \dots, r_d(\alpha))$ are the left and right eigenvectors for $\lambda(\alpha)$, normalized so that $\Sigma \ell_i r_i = 1$; $\Lambda^*(\cdot)$ = the convex conjugate of $\Lambda$; $Z_{ij}^n(\alpha) = \int e^{\alpha s} Z_{ij}^n(ds)$, $Z_i^n(\alpha) = (Z_{i1}^n(\alpha), \dots, Z_{id}^n(\alpha))$. $\langle \cdot, \cdot \rangle$ will denote the inner product.

**Hypotheses.**

(H1) $D = \{\alpha : \widehat{m}_{ij}(\alpha) < \infty$ for all $i, j\}$ contains a neighborhood of $O$; $M = \{m_{ij}\}$ is an irreducible aperiodic matrix;

(H2) $E\widehat{Z}_{ij}(\alpha)\widehat{Z}_{ik}(\alpha) < \infty$ for all $i, j, k$ and $\alpha \in \mathcal{D}$;

(H3) for fixed $a < \mu, \Lambda'(\alpha) = a$ has solution $\alpha_a$;
(H4) $\rho > 1$ (supercriticality);
(H5) $\underline{a}$ is taken to be an arbitrary but fixed point satisfying $\Lambda^*(a) < 0$.

Note that $\Lambda$ and $\Lambda^*$ look exactly like in Figures 2a and b, with $m$ replaced by $\rho$.

I will assume that (H1)–(H5) are operative throughout this section without further specific mention.

**Martingales.**

**Lemma 5.1.** *For all $\alpha \in \mathcal{D}$, $i = 1 \dots, d$*

$$W_i^n(\alpha) \equiv e^{-n\Lambda(\alpha)}\langle Z_i^n(\alpha), r(\alpha)\rangle \tag{5.1}$$

*is a martingale with respect to $\mathcal{F}_n$ = the $\sigma$-field generated by the entire history up to $n$ (including all information about particle production and motion).*

*Proof.* is by direct verification of the conditional expectation property. I omit the details.

**Behavior of the mean.**

**Lemma 5.2.**

$$\lim_{n\to\infty} \sqrt{n}e^{n\Lambda^*(a)}m_{ij}^n(an) = c(a)\ell_j(\alpha_a)r_i(\alpha_a), \tag{5.2}$$

*where $c(a)$ is a constant depending only on $\underline{a}$.*

This is a routine extension to the case of non-stochastic non-negative kernels of the result of Miller [34] for stochastic kernels $\{p_{ij}^n(\Gamma)\}$.

**Asymptotics of $\langle Z_i^n(an), r(\alpha_a)\rangle$.**

Since $W_i^n(\alpha) \geq 0$ is a martingale

$$\lim_{n\to\infty} W_i^n(\alpha) = W_i(\alpha), \qquad i = 1, \dots, d \tag{5.3}$$

exists w.p.1 for all $\alpha \in \mathcal{D}$. This can be seen by imitating the analogous argument for the multi-type Galton-Watson process (see [14]).

Due to hypothesis (H2), $EW(\alpha) = 1$.

**Theorem 5.1.** *Assume H(1) to H(5). Then*

$$\lim_{n\to\infty} \sqrt{n}e^{n\Lambda^*(a)}\langle Z_i^N(an), r(\alpha_a)\rangle = c(a)r_i(\alpha_a)W(\alpha_a) \quad a.s.. \tag{5.4}$$

*Note*: Constants like $c$ or $c(a)$ are generic, and need not be the same every time they appear.

The proof of the theorem is in the Tao's Thesis [40]. It follows closely the lines of Biggins corresponding result for the one type case [18]. Intuitively the result is natural: at the level of expectations it is (5.2) after taking inner products with $r(\alpha_a)$.

**Asymptotics of $Z_{ij}^n(an)$.**

The main new result in this note is the following:

**Theorem 5.2.** (Bramson,Ney,Tao) *Assume (H1-5). Then*

$$\lim_{n\to\infty} \sqrt{n}e^{n\Lambda^*(a)} Z_{ij}^n(an) = c(a)\ell_j(\alpha_a)r_i(\alpha_a)W(\alpha_a) \tag{5.5}$$

*in probability, for all $i, j = 1, \dots, d$.*

## Remarks

(1) The expectation version of (5.5) is just (5.2), but the jump from (5.2) to (5.5) is not trivial.
(2) At the logarithmic level, (5.5) just says

$$\frac{1}{n}\log Z_{ij}^n(an) \longrightarrow -\Lambda^*(a) \tag{5.6}$$

which is much easier to prove. The eigenvectors $\ell(\alpha_a)$ and $r(\alpha_a)$ are lost in (5.6). The sharper asymptotic form is thus needed for the following corollary.
(3) $Z_{ij}^n(an)$ in (5.5) can be replaced by the more general $Z_{ij}^n(an + o(n))$.

**Corollary.** *Under the hypothesis of theorem 5.2, on the set* $\{W(\alpha_a) > 0\}$,

$$\frac{Z_{ij}^n(an)}{Z_{hk}^n(an)} \longrightarrow \frac{\ell_j(\alpha_a)r_i(\alpha_a)}{\ell_k(\alpha_a)r_h(\alpha_a)} \tag{5.7}$$

*in probability, for* $i, j, h, k = 1, \ldots, d$.

In the ordinary multi-type G-W process

$$\frac{Z_{ij}^n}{Z_{hk}^n} \longrightarrow \frac{\ell_j}{\ell_k}, \quad \ell = \ell(0),$$

where $\ell M = \rho\ell$ (see e.g. [10] or [14]). Thus (5.7) shows how the population composition varies as $\underline{a}$ varies. (Note that $r_i(0) = 1$, $i = 1, \ldots, d$.)

**Outline of the proof.**

The details of the proof are somewhat lengthy and appear in [23]. The outline of the components of the proof is as follows: The behavior of $Z_{ij}^n(an)$ is studied in three parts.

(i) Show that all but a negligible number of particles that end up at $\underline{an}$ at time $n$, will have passed at time $n - N$ through a "small" ball centered at $a(n-N)$ for suitable $N$. The quantities $n, N$ and the size of the ball must be coordinated. (See Fig. 3.)
(ii) Show that the number of particles in the ball can be approximated in a suitable sense by its expectation.
(iii) Use an averaging argument to show that the offspring of the "large" number of particles in the "small" ball lead to the conjectured proportion of types at time $n$.

More precisely, let $Y_j^n(r, B)$ = the number of type $j$ particles at $\underline{an}$ at time $n$ which are descended from the $r$th particle in $B$ at time $N - n$, where $r = 1, \ldots, Z_i^{n-N}(B)$, and let $Y_j^n(B) = \sum_r Y_j(r, B)$. Write $IEY_j^n(B) = \sum_r (r, B)$ for the "internal" expectation of $Y_j^n$ (this is a random variable), and $\overline{Y}_j^n = Y_j^n - IEY_j^n$ for the centering of $Y_j^n$. (The index $i$ of the ancestor is suppressed in the $Y$'s.) Now we start with the decomposition

$$Z_{ij}^n(an) = IEY_j^n(B) + \overline{Y}_j^n(B) + Y_j^n(B_c). \tag{5.8}$$

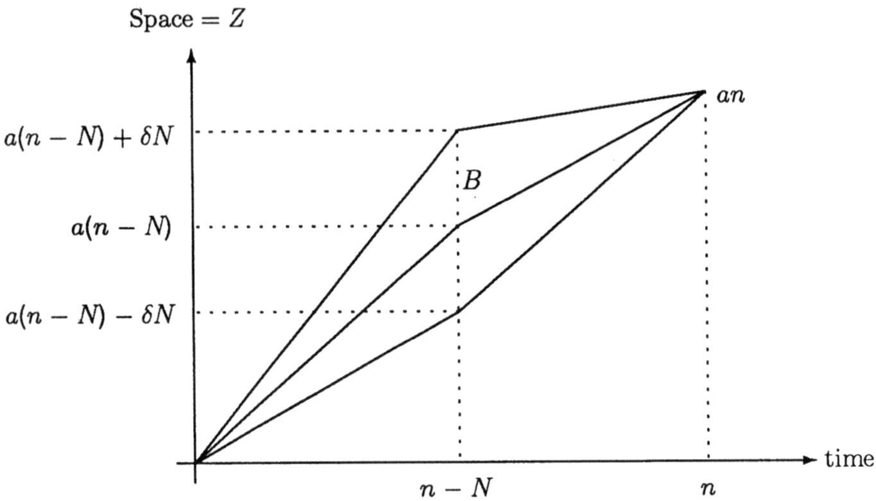


Figure 3

The random variables on the right side of this expression are compared to the rate function

$$\alpha(n, N, \Delta) = \alpha(n, N) = [N(n-N)]^{-1/2} \exp\{-n\Lambda^*(a) - N\Delta\},$$

with $B$ a ball of radius $\delta N$ centered at $a(n-N)$, and $n, N, \delta$ and $\Delta$ suitably coordinated. Namely we prove that

**Lemma 5.3.** *Given any $\delta > 0$ there exists a $\Delta_0 > 0$ such that*

$$\lim_{N\to\infty} \lim_{n\to\infty} \frac{EY_j^n(B^c)}{\alpha(n, N, \Delta_0)} = 0. \tag{5.9}$$

*This then implies that*

$$\lim_{n\to\infty} \frac{Y_j^n(B^c)}{\alpha(n, N)} \longrightarrow 0 \qquad \text{in probability} \tag{5.10}$$

*as $N \to \infty$.*

**Lemma 5.4.** *If $n \to \infty$ and $N \to \infty$ in such a way that $\frac{N}{n} \to 0$, then*

$$\frac{\overline{Y}_j(B)}{\alpha(n, N)} \longrightarrow 0 \quad \text{in probability.} \tag{5.11}$$

**Lemma 5.5.** *There exist a ball $B' \subset B$ of radius $\delta' < \delta$, and $\Delta' > 0$ and $\eta' > 0$ such that*

$$\liminf_{n\to\infty} \frac{IEY_j^n(B')}{\alpha(n, N, \Delta_0)} \geq \eta' e^{\Delta' N} W(\theta_a).$$

*This of course implies that for some $\eta > 0$*

$$\liminf_{n\to\infty} \frac{IEY_j^n(B)}{\alpha(n, N, \Delta_0)} \geq \eta W(\theta_a). \tag{5.12}$$

Of course similar results as in these lemmas hold with $Y^n$ replace by $\langle Y^n, r\rangle$. This fact together with (5.9)–(5.11) imply that

$$\left| \frac{Z_{ij}^n(an)}{\langle Z_i^n(an), r(\alpha_a)\rangle} - \frac{IEY_j^n(B)}{\langle IEY^n(B), r(\alpha_a)\rangle} \right| \longrightarrow 0 \tag{5.13}$$

in probability.

The proportions $\ell_j(\alpha_a)$ appear via the following "weak law":

**Lemma 5.6.** *Given any $\epsilon > 0$, there is a $\delta_0 > 0$ such that for sufficiently large $n$ and $N$, and for $\delta \leq \delta_0$*

$$\left| \frac{IEY_j^n(B)}{\langle IEY^n(B), r(\alpha_a)\rangle} - \ell_j(\alpha_a) \right| < \epsilon. \tag{5.14}$$

Finally (5.13) , (5.14) and Theorem 5.1 combine to yield Theorem 5.2.

## 6. Behavior outside of LD range

Since the eigenvectors $\ell(\alpha)$ and $r(\alpha)$ are positive for $\alpha_a \in \mathcal{D}$, the limit ratios in (5.7) cannot be 0, i.e. no type can be totally "dominant" over another at the point $an$. However, if for some $\underline{a}$ there is no solution of $\Lambda(\alpha) = a$ with $\alpha_a \in \mathcal{D}$, then the above analysis breaks down. Thus, for example, if the spatial motions are all $\geq 0$ and we consider $Z_{ij}^n(a)/Z_{ik}^n(a)$ (note the point $a$ is fixed, not $\underline{an}$) then by suitable choice of the spatial motion one might get dominance in the limit of some particle type, i.e. $Z_{ij}^n(a)/Z_{ik}^n(a) \to 0$.

This is what appears to be happening to the ratios of numbers of photons to electrons in the electron-photon cascade. In section 16.1 of chapter VII of Harris [10] some questions and conjectures related to this phenomenon are raised. Although these are outside the domain of our results in section 5, it does seem that we now have the tools to study these conjectures.

## References


[1] Harris, T. E., *Branching processes*, AMS **19** (1948), pp. 474–494.

[2] Harris, T. E., *Some mathematical models for branching processes*, in "2nd Berk. Symp.," 1951, pp. 305–328.

[3] Harris, T. E., *The random functions of cosmic-ray cascades*, PNAS **43** (1957), pp. 509–512.

[4] Harris, T. E., *A stationary measure for the multiplicative process*, Abstracts of short communication and scientific programme, in "International Congress of Mathematicians," Edinburgh, 1958.

[5] Harris, T. E., *A theorem on general branching processes*, AMS Notices **6** (1959a), p. 55.

[6] Harris, T. E., *A mathematical model for multiplication by binary fission*, in "The Kinetics of Cellular Proliferation," Grune and Stratton, Inc., New York, 1959b, pp. 368–381.

[7] Harris, T. E., *On one-dimensional neutron multiplication*, Research Memorandum RM-2317 (1960b), The RAND Corporation, Santa Monica, Calif..

[8] Harris, T. E., *Probability one convergence for age-dependent branching processes*, AMS Notices **7** (1960c), p. 486.

[9] Harris, T. E., *On the theory of age-dependent stochastic branching processes*, (with R. Bellman), Proc. Nat. Ac. Sc. **34** (1948), pp. 601–604.

[10] Harris, T. E., "The Theory of Branching Processes," Springer Verlag, Heidelberg, 1963.

[11] Asmussen, S. and Herring, H., "Branching Processes," Birkhauser, Boston, 1983.

[12] Asmussen, S. and Kaplan, N., *Branching random walks I and II*, J. of Stoch. Proc. and Appl. **4** (1976), pp. 1–32.

[13] Athreya, K. B. and Kaplan, N., *The additive property and its applications in branching processes*, Adv. in Prob. **5** (1978), pp. 27–60.

[14] Athreya, K. B. and Ney, P., "Branching Process," Springer Verlag, Heidelberg, 1972.

[15] Bahadur, R. R. and Rao, R., *On deviation of the sample mean,*, Ann. Math. Stat. **31** (1960), pp. 1015–1027.

[16] Bhabha, H. J. and Heitler, W., *The passage of fast electrons and the theory of cosmic showers*, PRS London A **159** (1937), pp. 432–458.

[17] Biggins, J. D., *Chernoff's theorem in branching random walk*, J. Appl. Prob. **14** (1977), pp. 630–636.

[18] Biggins, J. D., *Martingale convergence in the branching random walk*, J. of Appl. Prob. **14** (1977), pp. 25–37.

[19] Biggins, J. D., *The asymptotic shape of the branching random walk*, Adv. Appl. Prob. **10** (1978), pp. 62–84.

[20] Biggins, J. D., *Growth rates in the branching random walk*, Zeit. Wahr. **48** (1979), pp. 17–34.

[21] Bramson, M., *Maximal displacement of branching Brownian motion*, Comm. Pure and Appl. Math. **XXXI** (1978), pp. 531–581.

[22] Bramson, M., *Minimal displacement of branching random walk*, Zeit. Wahr. **45** (1978), pp. 89–108.

[23] Bramson, M., Ney, P. and Tao, J., *The composition of a spatial branching population*, Univ. of Wisconsin Technical Report (1989).

[24] Carleson, J. F. and Oppenheimer, J. R., *On multiplicative showers*, PR **51** (1937), pp. 220–231.

[25] Chauvin, B., *Arbres et processus de Bellman–Harris*, Ann. I. H. P. **22** (1986), pp. 209–232.

[26] Hammersley, J. M., *Postulates for subadditive processes*, Ann. Prob. **2** (1974), pp. 652–680.

[27] Janossy, L., *Note on the fluctuation problem of cascades*, Proc. Phys. Soc. (London) Sect. A, **34** (1950), pp. 241–249.

[28] Joffe, A., LeCam, L. and Neveu, J., *Sur la loi des grands nombres pour des variables aléatoires de Bernoulli attachées à un arbre dyadique*, C. R. Acad. des Sc. de Paris **277** (1973).

[29] Kingman, J. F. C., *The first birth problem for an age-dependent branching process*, Ann. Prob. **2** (1975), pp. 341–345.

[30] Kolmogorov, A. N., *Über das logarithmisch normale Verteilungsgesetz der Dimensionen der Teilchen bei Zerstükelung*, Doklady **31** (1941), pp. 99–101.

[31] Laredo, C. and Roualt, A., *Grandes deviations dynamique de populations, et phénomènes malthusiens,*, Ann. Inst. H. P. **4** (1983), pp. 323–350.

[32] Lopuszanski, J., *Some remarks on the asymptotic behavior of the cosmic ray cascade for large depth of the absorber: II. Asymptotic behavior of the probability distribution function*, Nuovo Cimento **X**, Ser. **2**, Suppl. **4** (1955), pp. 1150–1160.

[33] Lopuszanski, J., *Some remarks on the asymptotic behavior of the cosmic ray cascade for large depth of the absorber: III. Evaluation of the distribution function*, Nuovo Cimento **X**, Ser. **2**, Suppl. **4** (1955), pp. 1161–1167.

[34] Miller, H. D., *A convexity property in the theory of random variables on a finite Markov chain*, Ann. Math. Stat. **32** (1961), pp. 1260–70.

[35] Neveu, J., *Arbres et processus de Galton-Watson*, Ann. I. H. P. **22** (1986), pp. 199–208.

[36] Ney, P., *Generalized branching processes I & II*, Ill. J. of Math **8** (1964), pp. 316–349.

[37] Ney, P., *Ratio limit theorems for cascade processes*, Zeit. Wahr. **3** (1964), pp. 32–49.

[38] Ney, P., *The limit distribution of a binary cascade process*, J. of Math. Anal. and Appl. **10** (1965), pp. 30–36.

[39] Ney, P., *The convergence of a random distribution function associated with a branching process*, J. Math. Anal. and Appl. **12** (1965), pp. 316–327.

[40] Tao, J., Ph.D., *Thesis*, Univ. of Wisconsin, In preparation.

[41] Uchiyama, K., *Spatial growth of a branching process of particles living in* $\mathbf{R}^d$, Ann. Prob. **4** (1982), pp. 896–918.

[42] Urbanik, K., *Some remarks on the asymptotic behavior of the cosmic ray cascade for large depth of the aborber: I. Evaluation of the factorial moments*, Nuovo Cimento **X, Ser. 2, Suppl. 4** (1955), pp. 1147–1149.

[43] Watanabe, S., *Limit Theorems for a Class of Branching Processes*, in "Markov Processes and Potential Theory," Wiley, New York, 1967, pp. 205–32.

Department of Mathematics
University of Wisconsin
Madison, Wisconsin 53706

# Some Remarks on the Theory of Critical Branching Random Walk

J.T. Cox

*Abstract.* We give a brief review of the ergodic theory of critical branching random walk, and summarize some new work that extends this theory.

## 1. Introduction

Branching random walk is one of the simplest models for systems of particles which exhibit random motion and growth. The purpose of this note is to review the basic ergodic theory of the critical case of this model, and give a brief survey of some related results. We assume the particle motion takes place on the $d$-dimensional integer lattice $\mathbb{Z}^d$. $\eta_t$ will denote our model, a Markov process with state space $(\mathbb{Z}^+)^{\mathbb{Z}^d}$ (which we call configuration space). We interpret $\eta_t(x)$ as the number of particles at site $x$ at time $t$, $x \in \mathbb{Z}^d, t \geq 0$.

Here is a somewhat informal definition of our process. Let $q(x,y)$ be a random walk kernel on $\mathbb{Z}^d$. and except in Theorem 2 below, assume that

$$q(x,y) = 1(|x-y| = 1), \tag{1.1}$$

where $1_A = 1(A)$ is the indicator function of the set $A$. Thus $q(x,y)$ is the transition kernel for simple symmetric random walk. The dynamics are specified by requiring that all particles act independently of each other, and that at rate one each particle: (a) jumps according to $q(x,y)$, (b) dies, (c) splits into two. In (c), when the particle splits, the new particle is located at the site of the parent particle. We can rephrase this as follows. Each particle lives an exponentially distributed (with parameter two) lifetime, during which the particle performs rate one random walk with jump matrix $q(x,y)$. At the end of its lifetime a particle is replaced by zero or two particles, each possibility occuring with probability one-half.

Let $\eta_t^x$ denote the process starting from a single particle at $x$ at time 0. Observe that if we let

$$|\eta_t^x| = \sum_y \eta_t^x(y),$$

then $|\eta_t^x|$ is a nonspatial critical branching process, and thus $|\eta_t^x| = 0$ eventually for large $t$ with probability one. (Unrestricted sums are taken over all of $\mathbb{Z}^d$.) A construction of these "single ancestor" processes (in a quite general setting) can be found in Chapter 9 of Ethier and Kurtz (1986). With the single ancestor processes we can construct the general process by superposition. That is, let $\{\eta_t^{x,i}, x \in \mathbb{Z}^d, i = 1, 2, \dots\}$ be a family of independent branching random walks, each $\eta_t^{x,i}$ starting from a single particle at $x$ at time 0. Given any configuration $\zeta \in (\mathbb{Z}^+)^{\mathbb{Z}^d}$, we define a branching random walk $\eta_t$ with $\eta_0 = \zeta$ by letting

$$\eta_t(y) = \sum_x \sum_{i=1}^{\zeta(x)} \eta_t^x(y). \tag{1.2}$$

This construction shows that $\eta_t$ is an additive system in the sense of Harris (1978). Moreover, the construction works if $\zeta$ is random; we need only require that $\zeta$ and the $\eta_t^{x,i}$ be independent.

There are many related models. These include: models with more general branching/displacement mechanisms, discrete time models, "cluster" processes, branching Brownian motions and diffusions, measure-valued processes, etc. We consider only the very simple model defined above because it exhibits what is perhaps the most interesting phenomenon that the more general class of models displays.

To describe this phenomenon we need a little more notation. Let $\mathcal{L}$ denote law, and let $\Rightarrow$ denote weak convergence of probability measures. Thus the statement

$$\mathcal{L}(\eta_t) \Rightarrow \mu \text{ as } t \to \infty$$

means that for any finite $A \subset \mathbb{Z}^d$ and any configuration $\xi$,

$$\lim_{t\to\infty} P\big(\eta_t(x) = \xi(x) \text{ for all } x \in A\big) = \mu\big(\eta(x) = \xi(x)\big) \text{ for all } x \in A.$$

Finally, let 0 denote the configuration which is identically 0, and let $\delta_0$ be the unit point mass at 0. The basic ergodic theory of $\eta_t$ is given by the following result, which is a special case of several theorems in Durrett (1979).

**Theorem 0.** *Assume that $\mathcal{L}(\eta_0)$ is translation invariant on $\mathbb{Z}^d$, and that* $\theta = E[\eta_0(0)] < \infty$.

(A) *If $d \leq 2$ then as $t \to \infty$ ,*

$$\mathcal{L}(\eta_t) \Rightarrow \delta_0. \tag{1.3}$$

(B) *If $d \geq 3$ then there exists a one-parameter family $\{\nu_\theta, 0 \leq \theta < \infty\}$ of invariant measures for $\eta_t$ . The $\nu_\theta$ are translation invariant, and satisfy $\int \eta(0)d\nu_\theta(\eta) = \theta$. If, in addition, $\mathcal{L}(\eta_0)$ is shift ergodic on $\mathbb{Z}^d$, then as $t \to \infty$ ,*

$$\mathcal{L}(\eta_t) \Rightarrow \nu_\theta. \tag{1.4}$$

(A) above shows that in low dimensions ($d \leq 2$), starting from a "reasonable" initial state, for all $x \in \mathbb{Z}^d$,

$$\lim_{t\to\infty} P(\eta_t(x) = 0) = 0,$$

i.e. $\eta_t$ dies out as $t \to \infty$. This appears to be consistent with the fact that the single ancestor processes die out. On the other hand, (B) shows that in high dimensions ($d \geq 3$) $\eta_t$ is stable, i.e. there are nondegenerate equilibria. This seems a little surprising in light of the fact that the single ancestor processes die out. The key to understanding this is that particles somehow manage to "come in from infinity" to make up for this dying out.

For simple random walk the distinction between $d \leq 2$ and $d \geq 3$ is that of recurrence vs. transience. Kallenberg (1977) proved that this distinction for the underlying random walk kernel leads to the dichotomy in Theorem 0 for a class of discrete time models. Durrett (1979) proved similar results under some moment assumptions for models which include the one considered here.

We do not wish to pursue the intriguing question of transience vs. recurrence any further. Instead we wish to present a few results, including some work in progress, which extend Theorem 0 in some way. The first result is concerned with the approximation of infinite systems by large finite systems, and tries to capture the dichotomy between (A) and (B) for finite systems. The second result concerns the convergence in (B), and gives an improved convergence criteria. The third and last result addresses a question left open about invariant measures in (A). Namely, are there any invariant measures (apart from $\delta_0$) for $\eta_t$, necessarily of infinite mean?

## 2. Finite/Infinite Systems

Assume $d \geq 3$. We wish to consider the implications of part (B) of Theorem 0 for large finite systems. To do this, let $\Lambda_N = [-N, N)^d \cap \mathbb{Z}^d$, and let $q_N(x, y)$ be the random walk kernel $q(x, y)$ restricted to $\Lambda_N$ viewed as a torus. That is, for $x, y \in \Lambda_N$, let

$$q_N(x, y) = \sum_z q(x, z)1(z \equiv y \mod 2N).$$

We can define branching random walk $\eta_t^N$ on $\Lambda_N$ by using the same prescription as before, except that in (a) we use $q_N(x, y)$ instead of $q(x, y)$. Particles undergo splitting and death as before, but now perform random walk on $\Lambda_N$. We use $\eta_t^{N,x}$ to denote the process starting from a single particle at $x \in \Lambda_N$ at time 0.

We have two motives for investigating $\eta_t^N$. The first concerns the widely held assumption that the study of infinite particle systems like $\eta_t$ sheds light on the behavior of large finite systems (which are more realistic models for certain kinds of phenomena). We would like to make some quantitative statements confirming this assumption. The second motive takes the opposite point of view. That is, one can view the infinite systems as the primary objects of interest, and try to learn something about them through computer simulation of finite systems. In order to be confident that valid insights can be gained this way, it seems worthwhile to try to obtain some quantitative statements comparing the evolutions of finite and infinite systems.

The first thing to note in this program for branching random walk is that for fixed $N$, regardless of the initial state and dimension $d$, $|\eta_t^N|$ is a (nonspatial) critical branching process, and thus eventually dies out. The dichotomy of Theorem 0 is not evident for finite systems, however large. The problem is that we have been too crude by taking fixed $N$ and letting $t \to \infty$. If we allow both $N$ and $t$ to tend to infinity together we can do better, and indeed recover the distinction between low and high dimensions.

We need to introduce additional notation. For any configuration $\eta$ let

$$D_N(\eta) = (2N)^{-d} \sum_{x \in \Lambda_N} \eta(x)$$

be the average density of particles in $\eta$ on $\Lambda_N$. Let $I = [0, \infty)$ and let $Z_s$ be the diffusion on $I$ with transition function $Q_s$ and generator

$$\frac{1}{2} u \frac{d^2}{du^2}. \tag{2.1}$$

$Z_s$ is Feller's branching diffusion; 0 is an accessible absorbing boundary point, and starting from any point, $Z_s$ is eventually absorbed at 0 with probability one. We consider times $t_N$ such that as $N \to \infty$

$$t_N \to \infty \text{ and } \frac{t_N}{(2N)^d} \to s \in [0, \infty]. \tag{2.2}$$

The diffusion $Z_s$ and the finite systems $\eta_t^N$ are related by a well known fact. Let $\to_p$ denote convergence in probability. If $D_N(\eta_0^N) \to_p \theta$ as $N \to \infty$, and $t_N$ satisfies (2.2), then

$$D_N(\eta_{t_N}^N) \Rightarrow Z_s, Z_0 = \theta. \tag{2.3}$$

This result is well known and easy to derive: $Z_t^N = D_N(\eta_t^N)$ is a sequence of linear birth and death processes with generators converging to the one in (2.1).

To bring in the spatial structure of the $\eta_t^N$, we use the invariant measures $\nu_\theta$ from Theorem 0 (B) to define

$$\nu_\theta(s) = \int_I Q_s(\theta, d\theta')\nu_{\theta'}.$$

Since each $\nu_\theta(s)$ is a mixture of invariant measures, each $\nu_\theta(s)$ is invariant for the infinite system $\eta_t$. Note that $\nu_\theta(0) = \nu_\theta$ and that $\nu_\theta(\infty) = \delta_0$.

The last thing we must do before stating our comparison result is make some assumptions on the initial states $\eta_0^N$. We take a measure $\mu$ on the configuration space for the infinite system which is translation invariant, shift ergodic, and satisfies $\int \eta(0)d\mu(\eta) = \theta$. We construct $\eta_0$ with distribution $\mu$ and for each $N$ define $\eta_0^N$ by

$$\eta_0^N(x) = \eta_0(x) \text{ for } x \in \Lambda_N. \tag{2.4}$$

**Theorem 1.** *Assume that $d \geq 3$, and (2.2) and (2.4) hold. Then for any finite set $A \subset \mathbb{Z}^d$ and any configuration $\xi$,*

$$\lim_{N\to\infty} P\left(\eta_{t_N}^N(x) = \xi(x) \text{ for all } x \in A\right) = \nu_\theta(s)\left(\eta(x) = \xi(x) \text{ for all } x \in A\right). \tag{2.5}$$

Let us discuss Theorem 1 by examining some special cases of (2.5).

$s = 0$: In this case $t_N \to \infty$ but $t_N \ll N^d$. In view of (1.4) and (2.5) we can say that

$$\mathcal{L}(\eta^N_{t_N}) \approx \nu_\theta \approx \mathcal{L}(\eta_{t_N}).$$

For times $t$ which are large, but small compared to $N^d$, the distribution of $\eta^N_t$ is (locally) well described by the invariant measure $\nu_\theta$ of the infinite system.

$s = \infty$: In this case $t_N \gg N^d$, and (2.5) asserts that

$$\mathcal{L}(\eta^N_{t_N}) \approx \delta_0.$$

Roughly speaking, this shows that for $t$ large compared to $N^d$, $\eta^N_t$ has had time to realize it is a finite system, and dies out!

$0 < s < \infty$: In this intermediate case $t_N \approx s(2N)^d$. Keeping in mind (2.3), (2.5) says that the distribution of $\eta^N_{t_N}$ can be obtained by mixing the $\nu_{\theta'}$, with the mixing measure being (approximately) the distribution of $D_N(\eta^N_{t_N})$. This is also consistent with the $s = 0$ and $s = \infty$ cases. Roughly speaking, for $N$ and $t$ large, (2.3) and (2.5) suggest that the distribution of $\eta^N_t$ is approximately the invariant measure of the infinite system with intensity $D_N(\eta^N_t)$, i.e.

$$\mathcal{L}(\eta^N_t) \approx \nu_{D_N(\eta^N_t)}. \tag{2.6}$$

Theorem 1 and related results are proved in Cox and Greven (1990). Similar theorems for the voter model and the contact process are also discussed there.

## 3. Domains of attraction

In this section we will consider more general random walk kernels $q(x, y)$ defined on $\mathbb{Z}^d$. Namely, we assume throughout that

(3.1) $\qquad q(x, y)$ is irreducible, symmetric, and transient.

In this case the conclusions of Theorem 0 (B) are known to be true. A natural question to ask is whether it is possible to find more general conditions on the initial distribution $\eta_0$ which still insure the convergence in (1.4). That is, can one determine the domain of attraction of $\nu_\theta$?

There are several ways to approach (1.4). One method is given in Durrett (1979). Another is a method developed by Liggett and Spitzer (1981) developed for certain related processes. A third approach is to exploit the construction (1.2) and to make more or less "explicit calculations" with Laplace transforms. This approach, which gives a better criterion for convergence, is the one we will discuss here.

We start with some additional notation. Let $p_t(x,y)$, $t \geq 0$, denote the continuous time random walk kernel obtained by constructing a random walk with mean one exponential holding times at each state, and jump transition matrix $q(x,y)$. That is,

$$p_t(x,y) = e^{-t} \sum_{n=0}^{\infty} \frac{t^n}{n!} q^n(x,y).$$

**Theorem 2.** *Let $\eta_t$ be branching random walk with $q(x,y)$ satisfying (3.1). If for all $y \in \mathbb{Z}^d$*

$$\sum_x p_t(x,y)\eta_0(y) \to_p \theta \text{ as } t \to \infty, \tag{3.2}$$

*then $\mathcal{L}(\eta_t) \Rightarrow \nu_\theta$ .*

Note that moment assumptions and strict translation invariance are not required. An initial measure satisfying the assumptions of Theorem 0 (B) will automatically satisfy (3.2). We will make some general remarks concerning the method of proof of this result.

Fix a finite set $F \subset \mathbb{Z}^d$, and let $\alpha\colon F \to \mathbb{R}^+$, with $|\alpha| = \sum_{y\in F} \alpha(y) \geq 1$. Suppressing the dependence on $\alpha$ and $F$ in the notation, we define the transforms

$$\psi_\lambda(x,t) = E\Big(\exp\Big\{-\lambda \sum_{y\in F} \alpha(y)\eta_t^x(y)\Big\}\Big) - 1, \lambda \geq 0, t \geq 0, x \in \mathbb{Z}^d. \tag{3.3}$$

It is a routine matter to derive (using the generator, for instance) the equations

$$\psi_\lambda(x,t) = \sum_{y\in F} p_t(x,y)(e^{-\lambda\alpha(y)} - 1) + \int_0^t \sum_y p_{t-u}(x,y)[\psi_\lambda(y,u)]^2 du. \tag{3.4}$$

Using Jensen's inequality and the fact that $E(\eta_t^x(y)) = p_t(x,y)$, it is easy to see that

$$(3.5) \qquad |\psi_\lambda(x,t)| \le \lambda \sum_{y\in F} \alpha(y) p_t(x,y) \le \lambda |\alpha| p_t(0,0).$$

If we let $\mu = \mathcal{L}(\eta_0)$ , and use the basic construction (1.2), one can derive

$$(3.6) \quad E\Big[\exp\Big(-\lambda \sum_{y\in F} \alpha(y)\eta_t(y)\Big)\Big] = \int d\mu(\eta) \exp\Big(\sum_x \eta(x) \log(1+\psi_\lambda(x,t)\Big).$$

An important special case of the above is the case that the $\eta_0(x)$ are i.i.d. Poisson random variables with mean $\theta$. Then a calculation shows that the right-side of (3.6) reduces to

$$\exp\Big\{\theta \sum_x \psi_\lambda(x,t)\Big\}.$$

Using (3.4) this last sum becomes

$$\begin{aligned}\sum_{y\in F}(e^{-\lambda\alpha(y)}-1) + \int_0^t \sum_z [\psi_\lambda(z,u)]^2 du \\ \to \sum_{y\in F}(e^{-\lambda\alpha(y)}-1) + \int_0^\infty \sum_z [\psi_\lambda(z,u)]^2 du \\ \equiv \Psi(\lambda),\end{aligned}$$

where $\Psi$ depends on $\alpha$ and $F$. The assumptions on $q(x,y)$ can be used to show that $\Psi(\lambda)$ is finite; indeed the measures $\nu_\theta$ are characterized by

$$\int \exp\Big\{-\lambda \sum_y \alpha(y)\eta(y)\Big\} d\nu_\theta(\eta) = \exp \theta\Psi(\lambda).$$

Returning to (3.6), since (3.5) shows that $\psi_\lambda(x,t) \to 0$ uniformly as $t\to\infty$, it follows that $\log(1+\psi_\lambda(x,t)) \sim \psi_\lambda(x,t)$. Thus for the proof of Theorem 2 it suffices to prove that

$$(3.6) \qquad \sum_x \eta(x)\psi_\lambda(x,t) \to \theta\Psi(\lambda) \text{ in } \mu\text{-probability as } t\to\infty.$$

The previous paragraph and assumption (3.2) should make (3.7) at least plausible; the details of this argument are given in Cox (1990). As Maury Bramson says, it's only averaging.

## 4. Infinite mean invariant measures

We assume throughout this section that $d \leq 2$, and that $q(x, y)$ is the kernel for simple symmetric random walk given in (1.1). Theorem 0 (A) rules out the existence of any translation invariant measures with finite intensity which are invariant for $\eta_t$. That is, if $\mu$ is translation invariant, $\int \eta(0) d\mu(\eta) < \infty$, and $\mu$ is invariant for $\eta_t$, then Theorem 1 implies that $\mu = \delta_0$. This leaves open the question of the existence of invariant measure $\mu$ with $\int \eta(0) d\mu(\eta) = \infty$ (pointed out by Holley and Liggett (1981), and others). The subject of this section is: are there any invariant measure for $\eta_t$ other than $\delta_0$?

The analogous question for certain measure-valued diffusions was considered in Dawson et al. (1986). Although a definitive answer was not obtained, a very interesting convergence theorem was proved. The result suggests that the following should be true in our setting. Assume that $d = 1$, and the $\eta_0(x)$ are i.i.d. and

$$P(\eta_0(0) > n) \sim n^{-\alpha}. \tag{4.1}$$

Then

$$\cdot\ \mathcal{L}(\eta_t) \Rightarrow \begin{cases} \delta_0, & \alpha > 1/2 \\ \delta_\infty, & \alpha < 1/2 \\ c\delta_0 + (1-c)\delta_\infty, & \alpha = 1/2. \end{cases} \tag{4.2}$$

The meaning of $\Rightarrow \delta_\infty$ is this: for every $x \in \mathbb{Z}^d$ and $K < \infty$,

$$P(\eta_t(x) \leq K) \to 0 \text{ as } t \to \infty.$$

Thus (4.2) says that $\eta_t$ *dies out* if $\alpha > \frac{1}{2}$, *explodes* if $\alpha < \frac{1}{2}$, and tries to do *both* if $\alpha = \frac{1}{2}$ (i.e. $\eta_t$ dies out with probability $c$ and explodes with probability $(1-c)$). Note that $\delta_\infty$ is *not* a proper probability measure on configuration space.

(4.2) does not assert that $\delta_0$ is the only (proper) invariant measure for $\eta_t$, but it does show that if there is another invariant measure, no initial distribution of the type (4.1) can be in its domain of attraction. This suggests that there can be no invariant measure other than the trivial one. Cox and Greven (199?) have made some progress towards proving this. Note that in the critical case $\alpha = \frac{1}{2}$ above, there are on the order of $N^2$ particles within distance $N$ of the origin at time 0. In view of this fact, Cox and Greven make the following

**Conjecture.** *Assume $d = 1$, $q(x,y)$ is the transition kernel of simple symmetric random walk, and*

$$\kappa = \lim_{\delta \downarrow 0} \limsup_{N\to\infty} P(N^{-2} \sum_{x=-N}^{N} \eta_0(x) \geq \delta).$$

(i) *If $\kappa = 0$ then $\mathcal{L}(\eta_t) \Rightarrow \delta_0$ as $t \to \infty$.*
(ii) *If $\kappa > 0$ then there exists some $\epsilon > 0$ such that for all $K < \infty$ ,*

$$\limsup_{t\to\infty} P(\eta_t(x) \geq K) \geq \epsilon.$$

Given this, it is easy to see that there can be no invariant measures other than $\delta_0$. For suppose $\mu$ is invariant, $\eta_0$ has distribution $\mu$, and $\kappa = \kappa(\mu)$. If $\kappa = 0$ then (i) and the fact that $\mathcal{L}(\eta_t) = \mathcal{L}(\eta_0) = \mu$ implies $\mu = \delta_0$. On the other hand, if $\kappa > 0$, then (ii) and the fact that $P(\eta_t(x) \geq K) = P(\eta_0(x) \geq K) = \mu(\eta(x) \geq K)$ implies $\mu$ is not a proper probability measure.

The $d = 2$ case is more delicate, but the final conclusions should be the same as in the $d = 1$ situation. Hopefully, these matters will be resolved in Cox and Greven (199?).

## References


[1] J.T. Cox, *On the ergodic theory of critical branching Markov chains*, preprint (1991).

[2] J.T. Cox and A. Greven, *On the ergodic theory of critical branching Brownian motion in low dimensions*, preprint (1991).

[3] J.T. Cox and A. Greven, *On the long term behavior of some finite particle systems*, Prob. Th. Rel. Fields **85** (1990), pp. 195–237.

[4] D.A. Dawson, R.D. Foley, K. Fleischmann, L.A. Peletier, *A critical measure-valued branching process with infinite mean*, Stoch. Anal. Appl. **4** (1988), pp. 117–129.

[5] S.Ethier and T. Kurtz, "Markov Processes - Characterization and Convergence," Wiley, New York, 1986.

[6] R Durrett, *An infinite particle system with additive interactions*, Adv. Appl. Prob. **11** (1979), pp. 355–383.

[7] T.E. Harris, *Additive set-valued Markov processes*, Ann. Probab **6** (1978), pp. 355–378.

[8] R. Holley and T.M. Liggett, *Generalized potlatch and smoothing*, Z. Wahrsch. Verw. Gebeite **55** (1981), pp. 165–195.

[9] O. Kallenberg, *Stability of critical cluster fields*, Math. Nachr. **77** (1977), pp. 7–43.

[10] T.M. Liggett and F.L. Spitzer, *Ergodic theorems for coupled random walks and other systems with locally interacting components*, Z. Wahrsch. Verw. Gebeite **56** (1981), pp. 443–468.


Research supported in part by the NSF

Mathematics Department
Syracuse University
Syracuse, New York 13244

# Percolation

# Dynamic Renormalization and Continuity of the Percolation Transition in Orthants

David J. Barsky, Geoffrey R. Grimmett and Charles M. Newman

*Abstract.* A dynamic renormalization procedure, presented in the context of independent nearest-neighbor percolation in the $d$-dimensional orthant $\mathbb{Z}_+^d$, implies the absence of percolation at the critical point, together with related results.

## 0. Preface

It is a great pleasure to present a paper on percolation theory in honor of Ted Harris. Ted's only paper devoted to the subject [Ha1] was rather brief but fundamental, and appeared shortly after percolation was invented by Broadbent and Hammersley [BrH]. As is usual for a Harris paper it contained much of lasting value, preempting many of the techniques "discovered" later to be valuable for percolation in two dimensions. We list three items from that paper which will be touched upon in the ensuing pages:

(a) Lemma 4.1 of [Ha1]. These inequalities of Harris for independent random variables were rediscovered and put to good use in reliability theory (see [BP]) where they were later generalized to certain dependent systems [S]. The extension of Harris's inequalities in statistical mechanics [FKG] goes by the name of FKG inequalities.
(b) Remark of [Ha1]. Here Harris showed that the infinite cluster in two dimensions must be unique.
(c) Theorem 1 of [Ha1]. This theorem provides "half" of the proof that the two-dimensional (bond) model has a continuous phase transition.

In addition to these items, an important theme of Harris's paper which will play a major role in our work is the interplay between percolation in full-spaces, half-spaces, quadrants, etc.

## 1. Introduction

The purpose of this paper is two-fold. First, we report on the current status of continuity of the percolation phase transition and related issues concerning slabs and the uniqueness of the infinite cluster. This will be done in the next section, where we shall see that, for general dimensions, continuity has been proved for half-spaces, quarter-spaces, $\cdots$, and orthants [BaGN], but not yet for full-spaces. (On the other hand, the analogous issue for the contact process has been completely resolved [BezG].) Second, we present a dynamic renormalization technique, which leads to these continuity results. In order to minimize technical complications, we only present in this paper the version of the technique which is applicable to orthants. Those readers who feel deprived of complications can find a version applicable to half-spaces in [BaGN]. The remainder of this section provides an overview of our renormalization procedure.

Suppose one wishes to show that at a certain value $p_1$ of the bond density (i.e., the probability of any edge to be "open") for independent nearest-neighbor percolation on $\mathbb{Z}^3$, there exists percolation in some quarter-slice, $\mathbb{Q}_L \equiv \{-L, \cdots, L\} \times \mathbb{Z}_+^2$ (where $\mathbb{Z}_+ = \{0, 1, 2, \dots\}$). Partition $\mathbb{Q}_L$ into cubes which are translates of $\{-L, \cdots, L\}^3$ and label each cube with a site $x$ in the "renormalized" lattice, $\mathbb{L}' = \mathbb{Z}_+^2$, in the natural way. This also partitions the nearest-neighbor bonds of $\mathbb{Q}_L$ provided that bonds are allocated uniquely to cubes. One allocation method is to consider each cube as including those bonds leading out from its east and north faces; i.e., those faces having normal vectors (0,1,0) and (0,0,1). The structure of a conventional renormalization procedure (see, e.g., [R1, ACCFR]) is roughly as follows. Define the cube $C_x$ to be *occupied* if an event $A_x$ occurs, depending only on the bonds of $C_x$, which indicates that $C_x$ is intraconnected in some appropriate sense. The $A_x$'s will generally be translations of each other and should be chosen so that:

(i) Site percolation of occupied cubes in $\mathbb{L}'$ implies bond percolation in the quarter-slice $\mathbb{Q}_L$ of the original lattice.

(ii) For some $L$, the probability, $P_{p_1}(A_x)$, which represents the renormalized site density, exceeds $\lambda_c(\mathbb{Z}_+^2)$, the critical density for independent site percolation on $\mathbb{Z}_+^2$. In practice, this is typically shown by proving that for the given unrenormalized density $p_1$, $P_{p_1}(A_x)$ can be driven towards 1 by taking $L$ large.

The difficulty with this static renormalization procedure is that criterion (i) demands such a strong notion of intraconnectedness that criterion (ii) becomes impossible to verify when, for example, the only hypothesis on $p_1$ is that it is large enough to ensure percolation in the orthant $\mathbb{Z}^3_+$.

Our way of avoiding this difficulty is to develop a dynamic renormalization in which criterion (i) is automatically satisfied even with a fairly weak notion of intraconnectedness. This leads us to define connection events with a more complex structure. They are not translates of each other, and each depends not only on the bonds of the individual cube in question but also on information coming via the adjoining face of a neighboring cube. The renormalized model is still site percolation on $\mathbb{L}'$ (albeit dependent), but it is analyzed by a dynamic (or algorithmic) construction of the cluster of the origin, such as one might perform in a computer simulation.

In this construction, a site (i.e., a cube in $\mathbb{Q}_L$), other than the initial site, does not have its occupation status tested until some neighbor of that site has been found to be occupied. Thus any site found to be occupied will automatically belong to the cluster of the initial site. To decide which site to test at a given stage of the process, one utilizes a preassigned deterministic ordering of all sites in $\mathbb{Z}^2_+$. Then from the set of untested neighbors of sites already found to be occupied, the new site to be tested is chosen as the one with the lowest order. The site tested at stage $n$ is random but depends only on the past history of the running of the algorithm. If at some stage the set of untested neighbors is empty, then the algorithm terminates. Here criterion (ii) is replaced by

(ii′) For some $L$, the conditional probability that $C_x$ is occupied, given that it is tested, exceeds $\lambda_c(\mathbb{Z}^2_+) + \epsilon$ for some $\epsilon > 0$ and all $x \neq 0$.

The definition of *occupation* is as follows (where we no longer associate the bonds between adjoining faces of two cubes to one of the cubes in a deterministic way). For a fixed integer $M$ (chosen to depend on $L$ so as to make (ii′) valid), the cube at the origin $C_0$ is said to be *occupied* if the origin in $\mathbb{Q}_L$ is connected within $C_0$ to at least $M$ sites (which we call *active*) on the east and north faces of $C_0$. When another cube $C_x$ is tested, after one or more of its neighboring cubes have been found to be occupied, one first chooses one of the occupied neighbors (the one with the lowest preassigned order), say $C_{x'}$. Then $C_x$ is defined to be occupied if at least one of the $M$ or more active sites on the adjoining face of $C_{x'}$ is connected using the bonds only of $C_x$ (and those between the adjoining faces of $C_{x'}$ and $C_x$) to $M$ or more sites (again called active) on *each* of the three faces of $C_x$ (in the east-north plane) not adjoining $C_{x'}$.

The dynamic renormalization procedure just outlined leads immediately to Proposition 1 of Section 3, which, together with Proposition 2 of that section, implies continuity of the orthant transition. We remark that renormalization arguments with dynamical aspects related to those of our approach have appeared previously in the context of long range percolation [AN2] and directed percolation [BrDS, D]. Before continuing in Section 3 with the rest of the renormalization story, we present in the next section our main results together with background material and related results. A general account of the mathematics of percolation theory is to be found in [Gr].

## 2. Absence of Percolation at Criticality, and Related Results

Consider the standard $d$-dimensional bond percolation model; i.e., independent nearest-neighbor percolation on $\mathbb{Z}^d$. This is a random graph whose vertex set consists of all sites in $\mathbb{Z}^d$ and whose edge set is a randomly chosen subset of $\mathcal{B}_d$, the set of all nearest neighbor pairs in $\mathbb{Z}^d$. This random subset is the set of "open" edges in $\mathcal{B}_d$ obtained by declaring edges open (with probability $p$) or closed (with probability $1-p$) independently of each other. The parameter $p$ is called the bond density and $P_p$ (or just $P$) will denote the ensuing probability measure. The connected components of this random graph are called *clusters*, and the *percolation probability* is defined as

$$\theta(p) = P_p(\text{the cluster of the origin is infinite}).$$

Percolation is said to occur when $\theta(p) > 0$, or equivalently when there exists some infinite cluster with probability one (w.p.1). The existence of a percolation phase transition was proved by Broadbent and Hammersley:

**Theorem 1.** [BrH, H1, H2] *For $d \geq 2$, there exists $p_c$ in $(0,1)$ such that*

$$\theta(p) = 0 \textit{ for } p < p_c \quad \textit{and} \quad \theta(p) > 0 \textit{ for } p > p_c.$$

The main object of study in this paper is the value of $\theta(p_c)$. Before discussing that issue, we make some remarks about the continuity of $\theta$ and the uniqueness of the infinite cluster. It is easy to see that $\theta$ is always right-continuous; this is because it is the decreasing limit as $L \to \infty$ of the continuous functions,

$$P_p(\text{the cluster of the origin reaches the surface of } \{-L, \cdots, L\}^d).$$

It was then shown in [BeK] that for $p > p_c$, $\theta$ is continuous at $p$ if the infinite cluster is unique (w.p.1) at the density $p$. Uniqueness for all $p$ was first proved for $d = 2$ in [Ha1] and then for general $d$ in [AKN]. The best proof of uniqueness may be found in [BuK]; more recent developments are in [GKN]. Combining all these results, we have

**Theorem 2.** [Ha1, BeK, AKN] *For $d \geq 2$, $\theta$ is continuous for all $p$ in $[0, 1]$ if and only if $\theta(p_c) = 0$.*

Our motivation for caring about the value of $\theta(p_c)$ is partly because of Theorem 2 and partly because, in coming to understand the behavior of $\theta$ near $p_c$, one is led to develop new renormalization methods. We remark that there are certain independent percolation models in which $\theta(p_c) > 0$; these include long range percolation in one dimension [AN2], and nearest neighbor percolation in logarithmic wedges of $\mathbb{Z}^2$ [CC1]. It is generally believed that for standard percolation in $\mathbb{Z}^d$, $\theta(p_c) = 0$ for any $d$. However, such a result has only been proved for $d = 2$ and for large $d$ as indicated in the next two theorems. Following them, we explain how near we come to such a result for general $d$ by renormalization methods similar to those described in this paper.

**Theorem 3.** *(i)* [Ha1] *For $d = 2$, $\theta(\frac{1}{2}) = 0$. (ii)* [K1] *For $d = 2$, $p_c = \frac{1}{2}$.*

*Remark.* For other two-dimensional independent percolation models, such as site percolation in $\mathbb{Z}^2$, where $p_c$ is not explicitly known, $\theta(p_c)$ was proven to vanish by Russo using renormalization methods [R2].

**Theorem 4.** [HS1, HS2] *There exists $d_0 \geq 6$ such that $\theta(p_c) = 0$ if $d > d_0$.*

*Remark.* Hara and Slade [HS1,HS2] prove that the "triangle condition" is satisfied for $d > d_0$. This combined with other results [Ba,BaA] (see also [AKN]) implies $\theta(p_c) = 0$. The triangle condition also implies [AN1,Ba,N,BaA] that critical exponents take on mean field values; this cannot be true for $d < 6$ [T,CC2]. Hara and Slade give an upper bound of 92 on $d_0$ for nearest neighbor models; however, for certain other finite range models, they show that $d_0 = 6$ [HS1,HS2].

To state the main results of [BaGN], we consider bond percolation on certain subsets $\mathbb{L}$ of $\mathbb{Z}^d$ including the half-space

$$\mathsf{H} = \mathbb{Z}^{d-1} \times \mathbb{Z}_+,$$

the quarter-space, $\cdots$, the orthant

$$\mathbb{O} = \mathbb{Z}_+^d,$$

and for $d \geq 3$ also the slabs

$$\mathbb{S}_L = \mathbb{Z}^{d-1} \times \{0, \cdots, L\},$$

and quarter-slices

$$\mathbb{Q}_L = \{-L, \cdots, L\}^{d-2} \times \mathbb{Z}_+^2.$$

We denote by $\theta_{\mathbb{L}}(p)$ the probability that the origin belongs to an infinite cluster using only the vertices and edges in $\mathbb{L}$, and by $p_c(\mathbb{L})$ the critical density for percolation on $\mathbb{L}$. The next theorem is stated for all $d \geq 2$; the $d = 2$ case was already known following work of [Ha1, K1, R1, SW]. The identity between the limit of $\mathbb{S}_L$ and $\mathbb{Q}_L$ critical densities for $d \geq 3$ was also already known [K2].

**Theorem 5.** [BaGN] *Set $d \geq 2$. (i) For $\mathbb{L} = \mathbb{Z}^{d-e} \times \mathbb{Z}_+^e$ $(1 \leq e \leq d)$,*

$$\theta_{\mathbb{L}}(p_c(\mathbb{L})) = 0. \tag{1}$$

*(ii) The sets $\mathbb{L}$ of part (i) have a common critical density: $p_c(\mathbb{H}) = p_c(\mathbb{Z}^{d-2} \times \mathbb{Z}_+^2) = \cdots = p_c(\mathbb{O})$; for $d \geq 3$, it is also the case that*

$$p_c(\mathbb{H}) = \lim_{L\to\infty} p_c(\mathbb{S}_L) = \lim_{L\to\infty} p_c(\mathbb{Q}_L). \tag{2}$$

*Remark.* Combining (1) and (2) with the results and arguments of [ACCFR] shows that the infinite cluster in $\mathbb{Z}^{d-e} \times \mathbb{Z}_+^e$ $(1 \leq e \leq d)$ is unique. (This was already shown for $e = 1$ in [K2].) Making use of the uniqueness of the infinite cluster, together with the arguments of [BeK], we obtain from Theorem 5 the continuity of $\theta_{\mathbb{L}}(p)$ at all bond densities $p$ for the sets $\mathbb{L}$ dealt with in part (i) of the theorem. Note that this does not yield continuity for either the full-space $\mathbb{Z}^d$ or the slabs $\mathbb{S}_L$.

*Remark.* As a corollary of Theorem 5, it is immediate that a resolution of the following would imply $\theta(p_c) = 0$ (i.e., continuity of $\theta$ for $\mathbb{L} = \mathbb{Z}^d$):

$$\text{prove that for any } p, \quad \theta(p) > 0 \text{ implies } \theta_{\mathbb{H}}(p) > 0. \tag{3}$$

This open problem is somewhat stronger than the statement that $p_c(\mathbb{Z}^d) = p_c(\mathbf{H})$, which is not itself a consequence of the above theorem. However, Grimmett and Marstrand [GrM] have recently proven that $p_c(\mathbb{Z}^d) = p_c(\mathbf{H})$ by an extension of the dynamic renormalization procedure of [BaCN]. This last result, combined with (2), verifies the identity, $p_c(\mathbb{Z}^d) = \lim_{L\to\infty} p_c(\mathbb{S}_L)$, for $d \geq 3$ as originally conjectured in [ACCFR]. This identity and the similar one for quarter-slices instead of slabs have various consequences concerning supercritical behavior in $\mathbb{Z}^d$ (see [GrM]). The Grimmett-Marstrand result also shows that (3) is necessary as well as sufficient for continuity of $\theta$ in $\mathbb{Z}^d$.

The results of Theorem 5 for $d = 2$ were already known. For $d \geq 3$, they follow immediately from the next theorem, whose proof is given in [BaGN] based on a renormalization procedure similar to, but more complicated than, the one presented in the next section. (A version of the [BaGN] arguments can also be applied when $d = 2$.)

**Theorem 6.** [BaGN] *Set* $d \geq 3$. *For any* $p$ *in* $(0,1)$,

$$\theta_{\mathbf{H}}(p) > 0 \quad \textit{implies} \quad \theta_{\mathbb{Q}_L}(p-\epsilon) > 0$$

*for some* $L < \infty$ *and* $\epsilon > 0$.

Although Theorems 5 and 6 do not quite resolve the issue of (non)percolation at the critical point for percolation on $\mathbb{Z}^d$, an extension of the renormalization methods of [BaGN] has been used by Bezuidenhout and Grimmett to resolve the analogous issue of (non)survival of the contact process at its critical point [BezG]. The contact process is an interacting particle system model with spatial coordinate in $\mathbb{Z}^{d-1}$ and time coordinate in $[0,\infty)$ (or $\mathbb{Z}_+$) which, as originally realized by Harris [Ha2], is closely related to directed (or oriented) percolation on $\mathbb{Z}^{d-1} \times \mathbb{Z}_+$. The analogue of Theorem 6 states that, for $d \geq 2$, if the process on $\mathbb{Z}^{d-1} \times [0,\infty)$ survives with infection rate $\lambda$ (the analogue of $p$), then for some $L$ and $\epsilon$ it survives on $\{-L,\cdots,L\}^{d-2} \times \mathbb{Z} \times [0,\infty)$ with rate $\lambda - \epsilon$.

We close this section by expressing the (perhaps pious) hope that further extensions of the dynamic renormalization procedure may yield further results beyond those of [BaGN, GrM, BezG], including proof of the continuity of the percolation transition in $\mathbb{Z}^d$ for general $d$.

## 3. The Renormalization Proof

In this section we present a proof, based on the type of dynamic renormalization procedure discussed in Section 1, of the following weakened version of Theorem 6.

**Theorem 6′.** *Set $d \geq 3$. For any $p$ in $(0,1)$,*

$$\theta_{\mathbb{O}}(p) > 0 \quad \textit{implies} \quad \theta_{\mathbb{Q}_L}(p-\epsilon) > 0$$

*for some $L < \infty$ and $\epsilon > 0$.*

Theorem 6′ is an easy consequence of the following two propositions. For the first proposition, we need some definitions. For given positive integers $L$ and $M$ ($M \leq (2L+1)^{d-1}$), let $T_{L,M}$ denote the collection of subsets $t$ of $F \equiv \{-L, \cdots, L\}^{d-1} \times \{-1\}$ containing exactly $M$ points. Think of $F$ as the north face of the $(2L+1)^d$-cube $C'$ below (to the south of) the cube $C = \{-L, \cdots, L\}^{d-1} \times \{0, \ldots, 2L\}$. The cube $C$ has south, north, west and east faces:

$$\begin{aligned}
F_s &= \{-L, \cdots, L\}^{d-1} \times \{0\}, \\
F_n &= \{-L, \cdots, L\}^{d-1} \times \{2L\}, \\
F_w &= \{-L, \cdots, L\}^{d-2} \times \{-L\} \times \{0, \cdots, 2L\}, \\
F_e &= \{-L, \cdots, L\}^{d-2} \times \{L\} \times \{0, \cdots, 2L\}.
\end{aligned}$$

Define $\mathcal{B}$ to be the set of (nearest neighbor) edges either completely within $C$ or else between $C'$ and $C$ (i.e., between adjacent sites in the face $F$ of $C'$ and the face $F_s$ of $C$). Define for $t \in T_{L,M}$

$$\begin{aligned}
R_{L,t}(p) = P_p(&\text{at least one site in } t \text{ is connected by open edges in } \mathcal{B} \\
&\text{to } M \text{ or more sites in } \textit{each} \text{ of the three faces, } F_n, F_w \\
&\text{and } F_e)
\end{aligned}$$

and then define

$$R_{L,M}(p) = \min_{t \in T_{L,M}} R_{L,t}(p).$$

Note that $R_{L,M}$ is a continuous function of $p$ since it is a "finite volume" quantity.

**Proposition 1.** *Set $d \geq 3$. Let $\lambda_c(\mathbb{Z}_+^2)$ denote the critical density of independent nearest neighbor site percolation in the quadrant $\mathbb{Z}_+^2$. Suppose $p$ is such that for some $L$ and $M$,*

$$R_{L,M}(p) > \lambda_c(\mathbb{Z}_+^2).$$

*Then for the same $p$ and $L$, $\theta_{\mathbb{Q}_L}(p) > 0$; i.e., bond percolation occurs in the quarter slab $\mathbb{Q}_L$ at bond density $p$.*

*Proof.* This is a fairly immediate consequence of the renormalization procedure presented in Section 1 above. The point is to consider the random set of all sites in the renormalized lattice $\mathbb{Z}_+^2$ (each of which corresponds to a $(2L+1)^d$-cube from $\mathbb{Q}_L$) eventually found to be occupied by the dynamic renormalization process. It suffices to show that this random set stochastically dominates (under the inclusion ordering) the random cluster of the origin constructed from an independent site percolation model on $\mathbb{Z}_+^2$ in which the site $x$ is occupied with probability $\mu_x$ where

$$\mu_x = \begin{cases} R_{L,M}(p), & \text{for each } x \neq 0, \\ P_p \ \text{(the cube } C_0 \text{ is occupied)}, & \text{for } x = 0. \end{cases}$$

To see why this is so, extend the algorithmic construction as follows. Let $W_1$, $W_2, \cdots$, be i.i.d. random variables uniformly distributed on [0,1] which are independent of our original bond percolation model. Let $x_n$ be the (random) $n$-th site in $\mathbb{Z}_+^2$ (excluding the origin) to be tested by the algorithm and let $\lambda_n$ be the (random) conditional probability that $x_n$ is occupied given the past history of the process. Declare the origin to be *red* if it is occupied and declare each $x_n$ to be red if it is occupied *and* $W_n < R_{L,M}(p)/\lambda_n$. Then the connected component containing the origin of the set of all sites eventually found to be red provides a realization of the cluster of the origin for independent site percolation with occupation probabilities $\{\mu_x\}$. This random red cluster is contained in, hence dominated by, the set of all sites eventually found to be occupied. Q.E.D.

**Proposition 2.** *Set $d \geq 2$. If, for some $p$ in $(0,1)$, $\theta_{\mathbf{O}}(p) > 0$, then for any $\delta > 0$, there exist some $L$ and $M$ such that $R_{L,M}(p) \geq 1 - \delta$.*

Before giving the proof of Prop.2, which is the last and largest part of this section, we give the much easier derivation of Theorem 6′ from the above two propositions.

*Proof of Theorem 6′ (from Props. 1 and 2).* Suppose $\theta_{\mathbb{O}}(p) > 0$. By Prop. 2, we choose $L$ and $M$ so that $R_{L,M}(p) > \lambda_c(\mathbb{Z}_+^2)$. But $R_{L,M}$ is continuous, and so for some $\epsilon > 0$, $R_{L,M}(p-\epsilon) > \lambda_c(\mathbb{Z}_+^2)$. Then by Prop. 1, $\theta_{\mathbb{Q}_L}(p-\epsilon) > 0$ and the proof is complete. Q.E.D.

We now begin our proof of Prop. 2. The basic building block of the proof is the next lemma. For ease of presentation, we henceforth restrict attention to $d = 3$. We denote by $C^K$ the cube $\{0, \cdots, K\}^3$ and we denote its north, east and up faces (the three faces of $C^K$ which do not contain the origin) by

$$F'_n = \{0, \cdots, K\}^2 \times \{K\},$$
$$F'_e = \{0, \cdots, K\} \times \{K\} \times \{0, \cdots, K\},$$
$$F'_u = \{K\} \times \{0, \cdots, K\}^2,$$

and similarly denote its south, west and down faces (the three faces of $C^K$ which do contain the origin) by $F'_s$, $F'_w$, $F'_d$. We also denote by $B_k$ (for $k \le K$), the size-$k$ "corner" of $C^K$,

$$B_k = \{x \in F'_s \cup F'_w \cup F'_d : \| x \|_\infty \le k\},$$

where $\| \cdot \|_\infty$ denotes the sup norm.

**Lemma 1.** *Suppose $\theta_{\mathbb{O}}(p) > 0$; then for any $\delta_1 > 0$, there exists some $k = k(\delta_1)$ such that for any $M'$ there is some $K_1 = K_1(M', k)$ such that*

$$V_{K,M',k}(p) \equiv P_p(B_k \textit{ is connected by open edges within } C^K \textit{ to } M' \textit{ or more sites in each of } F'_n,\ F'_e \textit{ and } F'_u)$$

*satisfies*

$$V_{K,M',k}(p) \ge 1 - \delta_1 \quad \textit{for all} \quad K \ge K_1.$$

*Proof.* For $i = n$ or $e$ or $u$, we can write

$$V_{K,M',k} = P(A_n \cap A_e \cap A_u)$$

where $A_i$ is the event that at least one site in $B_k$ is connected by open edges within $C^K$ to $M'$ or more sites in $F'_i$. Note that by symmetry the $A_i$'s have the same probability and are increasing events; thus we may use the Harris (FKG) inequalities [Ha1] as in [R1]:

$$\begin{aligned}1 - P(A_n \cup A_e \cup A_u) &= P(A_n^c \cap A_e^c \cap A_u^c) \geq P(A_n^c) \cdot P(A_e^c) \cdot P(A_u^c) \\ &= [P(A_i^c)]^3 = [1 - P(A_i)]^3,\end{aligned}$$

and thus

$$\begin{aligned}P(A_n \cap A_e \cap A_u) &\geq P(A_n) \cdot P(A_e) \cdot P(A_u) = [P(A_i)]^3 \\ &\geq [1 - (1 - P(A_n \cup A_e \cup A_u))^{1/3}]^3.\end{aligned}$$

Now $A_n \cup A_e \cup A_u$ contains the event that

$$\begin{aligned}X_{K,k} \equiv \text{ the number of sites in } & F_n' \cup F_e' \cup F_u' \\ \text{connected to } B_k \text{ by open edges within } & C^K\end{aligned}$$

is at least $3M'$. Thus to prove the lemma, it suffices to show that for any $M'$

$$\lim_{k\to\infty} \liminf_{K\to\infty} P(X_{K,k} \geq 3M') = 1.$$

Let us denote by $T_k$ the supremum of the $K$'s such that $X_{K,k} > 0$. Then $T_k = \infty$ if and only if some $x$ in $B_k$ belongs to an infinite cluster in $\mathbb{O}$, and clearly $P(T_k = K+1) \to 0$ as $K \to \infty$. But

$$\begin{aligned}P(T_k = K+1) &\geq \sum_{j=1}^{3M'-1} P(X_{K,k} = j) P(X_{K+1,k} = 0 | X_{K,k} = j) \\ &\geq (1-p)^{3(3M'-1)} P(0 < X_{K,k} < 3M').\end{aligned}$$

Thus

$$\liminf_{K\to\infty} P(X_{K,k} \geq 3M') = 1 - \lim_{K\to\infty} P(X_{K,k} = 0) = P(T_k = \infty).$$

But, by a simple ergodicity argument, some point in the boundary of $\mathbb{O}$ belongs to an infinite cluster in $\mathbb{O}$ (w.p.1) if $\theta_{\mathbb{O}}(p) > 0$, and hence $P(T_k = \infty) \to 1$ as $k \to \infty$ which completes the proof of Lemma 1. Q.E.D.

In the subsequent proof of Prop. 2, the connections across the size $L$ cube $C$ are constructed by attaching together several building blocks like those from Lemma 1 of smaller size(s) $K$. The attachments require some mortar at the corners of the building blocks. The mortar is provided by Lemmas 2 and 3. Lemma 2 will be used to reach the north face of $C$ and Lemma 3 (along with Lemma 2) to reach the west and east faces. In Lemma 2, we consider the "face"

$$F_h = \{-L, \cdots, L\}^{d-1} \times \{h\}$$

which for $h = L$ or $L-1$ is approximately halfway between the north face $F_n$ of $C$ and the north face $F$ of the adjoining cube $C'$. We also denote by $\mathcal{B}_{(h)}$ the set of those edges in $\mathcal{B}$ which do not go north of $F_h$.

**Lemma 2.** *Suppose $\theta_{\mathbb{O}}(p) > 0$ and $\delta_2 > 0$; then there exists some $M_2 = M_2(\delta_2)$ such that for any $M_2' \geq M_2$ there is some $K_2 = K_2(M_2', \delta_2)$ such that, for every $L \geq K_2$, $t \in T_{L,M_2'}$ and $h = L$ or $L-1$,*

$$P_p(\textit{at least one site in } t \textit{ is connected by open edges in } \mathcal{B}_{(h)} \textit{ to } M_2' \textit{ or more sites in } F_h) \geq 1 - \delta_2.$$

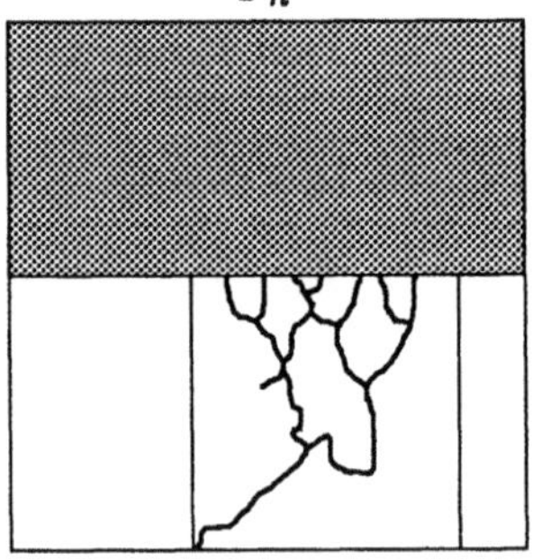


Figure 1.

A schematic diagram, in the $x_2$ - $x_3$ plane, of a three-dimensional crossing event occurring in the proof of Lemma 2. Here, a well-connected corner at $\overline{x}$, a site in $F_s$, the south face of the cube $C$, is connected within a cube emanating from $\overline{x}$ to many sites halfway to $F_n$, the north face of $C$.

*Proof.* (See Figure 1.) We begin with the $M_2'$ sites of some $t$ in the face $F$ of the cube $C'$ just below the south face $F_s$ of cube $C$. In the course of our discussion, we will obtain various lower bound restrictions on $M_2'$; $M_2$ will be chosen to satisfy all these restrictions. First we wish the $M_2'$ sites of $F$ to be connected (with high probability) to many (say $J_1$) sites in $F_s$ by the single north-south bonds between $F$ and $F_s$; for this purpose, we choose $M_2'$ big enough so that the probability of at least $J_1$ successes out of $M_2'$ Bernoulli ($p$) trials exceeds $1 - \delta_2/3$. Of the $J_1$ or more sites reached so far in $F_s$, we want to find a large number (say $J_2$) of sites $x$ in *one* of the four quarters of $F_s$ ($F_s \cap \{\pm x_1 \geq 0\} \cap \{\pm x_2 \geq 0\}$) with the property that the size-$k$ corners, $B_k(x) + x$, emanating from these sites do not overlap. Here, $B_k(x)$ is the corner $B_k$ but with its $x_1$ and/or $x_2$ coordinate reflected depending on which quarter of $F_s$ contains $x$:

$$B_k(x) = \begin{pmatrix} \mp 1 & 0 & 0 \\ 0 & \mp 1 & 0 \\ 0 & 0 & 1 \end{pmatrix} B_k.$$

The size $k$ is that given by Lemma 1 in the form $k(\delta_2/3)$ (i.e., with the $\delta_1$ of Lemma 1 taken to be $\delta_2/3$). The existence of $J_2$ such sites will be ensured if, say, $J_1/(4k+1)^2 \geq 4J_2$.

Without loss of generality, we suppose these $J_2$ sites are in the $x_1 \leq 0$, $x_2 \leq 0$ quarter of $F_s$. (For each of these sites, $B_k(x)$ is just $B_k$.) We wish to inspect the bonds within the $J_2$ corners, one corner at a time in the specific order described below, until we find one of the corners, say $B_k + \bar{x}$, which has all its sites connected to each other by open edges within the corner. We choose $J_2$ so large that the probability of finding at least one such corner is at least $1 - \delta_2/3$; this can be done by ensuring that the probability of one or more successes out of $J_2$ Bernoulli ($\tilde{p}_k$) trials exceeds $1 - \delta_2/3$, where $\tilde{p}_k = \tilde{p}_k(p, k) > 0$ is the probability that an individual size-$k$ corner is "well-connected".

We wish to apply Lemma 1 to obtain connections from $B_k + \bar{x}$ to many sites on the north face of $C^h + \bar{x}$ (which is within $F_h$). In order to do so, we need to choose an order of inspection so that when we find our well-connected corner $B_k + \bar{x}$, we have not already inspected any of the bonds in $C^h + \bar{x}$ during the earlier inspections. (The fact that we have inspected $B_k + \bar{x}$ itself before applying Lemma 1 will cause no difficulty.) It turns out that because the $(B_k + x)$'s to be inspected have already been chosen not to overlap, ordering the $x$'s by $x_1 + x_2$ (with some deterministic rule to break ties) gives an appropriate inspection order, as the reader may verify. Now Lemma 1 is applied as indicated with the $\delta_1$, $M'$ and $K$ of Lemma 1 taken to be $\delta_1 = \delta_2/3$, $M' = M_2'$, and $K = h = L$ or $L-1$. Since $K$ must be at least the $K_1$ of Lemma 1, we may essentially choose the $K_2(M_2', \delta_2)$ of Lemma 2 to equal $K_1(M_2', k(\delta_2/3))$. Q.E.D.

The next lemma concerns connections in a half-cube. Its proof uses both the arguments of Lemma 2, and Lemma 2 itself. We denote by $\tilde{\mathcal{B}}$ the set of those edges in $\mathcal{B}$ which do not go west of the plane $x_2 = 0$.

**Lemma 3.** *Suppose $\theta_{\mathbf{0}}(p) > 0$ and $\delta_3 > 0$; then there exists some $M_3 = M_3(\delta_3)$ such that for any $M_3' \geq M_3$ there is some $K_3 = K_3(M_3', \delta_3)$ so that for every $L \geq K_3$ and every subset $\tilde{t}$ of $\tilde{F} \equiv \{-L, \cdots, L\} \times \{0, \cdots, L\} \times \{-1\}$ containing exactly $M_3'$ sites,*

$$P_p(\textit{at least one site in } \tilde{t} \textit{ is connected by open edges in } \tilde{\mathcal{B}} \textit{ to } M_3' \textit{ or more sites in } F_e) \geq 1 - \delta_3.$$

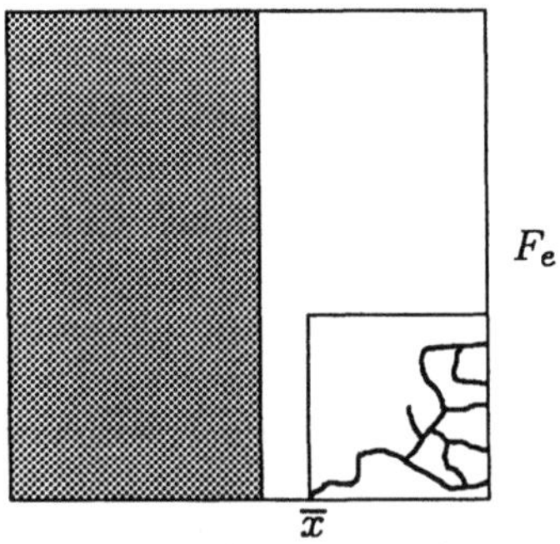


Figure 2.

In the proof of Lemma 3, a well-connected corner at $\overline{x}$, a site in $F_s$ which is *not* close to $F_e$, is connected within the eastern half of the cube $C$ to many sites in $F_e$.

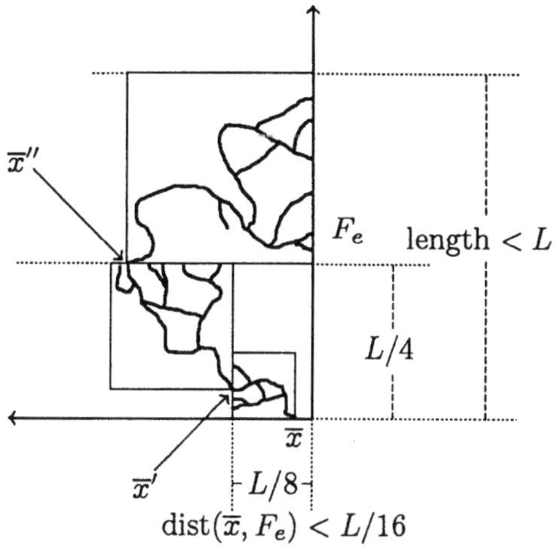


Figure 3.
A crossing event in the southeastern portion of the cube $C$, which occurs in the proof of Lemma 3. Here, a well-connected corner at $\overline{x}$, a site in $F_s$ which *is* close to $F_e$, is connected to many sites in $F_e$. The construction involves cubes emanating from $\overline{x}$ and two more well-connected corners at $\overline{x}'$ and $\overline{x}''$.

*Proof.* (See Figures 2 and 3.) We proceed as in Lemma 2, using single north-south bonds to connect the $M_3'$ sites in $\tilde{F}$ to many sites $x$ in $\tilde{F}_s \equiv \{-L, \cdots, L\} \times \{0, \cdots, L\} \times \{0\}$. Before testing any corners, we see whether many of these sites in $\tilde{F}$ are further than distance $L/16$ from the east face $F_e$ (i.e., have $L - x_2 > L/16$). If so, then (without loss of generality) we may assume that there are many which also have $x_1 \leq 0$, and we begin inspecting corners $B_k + x$ as in Lemma 2 until we find a well-connected one $B_k + \overline{x}$. We then apply Lemma 1 to obtain connections from $B_k + \overline{x}$ to many sites on the east face of $C^{L-\overline{x}_2} + \overline{x}$ (which is within $F_e$). See Figure 2.

If there are not many sites with $L - x_2 > L/16$, then there are many with $L - x_2 \leq L/16$ and (without loss of generality) with $x_1 \leq 0$. We now apply a reflected version ($x_2 \to -x_2$) of the arguments of Lemma 2 to obtain connections from a well-connected (reflected) corner at some $\overline{x}$ to many sites on the plane at $x_2 = L - L/8 + 1$. Repeating this type of argument, we find a well-connected corner at some $\overline{x}'$ on the plane $x_2 = L - L/8$ and then apply Lemma 1 to eventually reach a well-connected corner at some $\overline{x}''$ on the plane $x_3 = L/4$ (with $L - \overline{x}_2'' \geq L/8$); the latter step is possible since $\overline{x}_3' \leq L/8$. We finally apply Lemma 1 to obtain connections from $B_k + \overline{x}''$ to many sites on the east face of $C^{L-\overline{x}''_2} + \overline{x}''$ (which is within $F_e$). See Figure 3. Q.E.D.

We are now ready to combine Lemmas 2 and 3 to prove Prop. 2.

*Proof of Prop. 2.* Let $t \in T_{L,M}$ denote a particular set of sites in the face $F$ of $C'$. For $i = n$ or $w$ or $e$, the event $A_i$, that at least one site in $t$ is connected by open $\mathcal{B}$-edges to $M$ or more sites in $F_i$, is an increasing event. Thus by the Harris (FKG) inequalities [Ha1], it suffices to show separately for $i = n$ or $w$ or $e$, that $P_p(A_i)$ is close to 1 (for all $t$) for some choice of $L$ and $M$.

To show that $P(A_n)$ is close to 1, i.e., to obtain appropriate south to north crossings of $C$, we simply apply Lemma 2 twice. The first application reaches many sites in (say) the "face" $F_L$ at $x_3 = L$ and the second application gets from at least one of these sites to the north face $F_n$ at $x_3 = 2L$.

To show that $P(A_e)$ is close to 1 (and by symmetry, to do the same for $P(A_w)$), i.e., to obtain appropriate south to east crossings of $C$, we first see whether there are many sites of $t$ in the eastern half of thc facc $F$. If so, we can directly apply Lemma 3 to reach the east face $F_e$. If not, then there are many sites of $t$ in the western half of $F$ and we apply a translated (by $-L$ in the $x_2$ direction) version of Lemma 3 to reach many sites in the plane $x_2 = 0$. To continue from these to reach the east face, we apply (a rotated version) of Lemma 2. The proof is now complete. Q.E.D.

## References


[ACCFR] Aizenman, M., Chayes, J. T., Chayes L., Fröhlich, J. and Russo, L., *On a sharp transition from area law to perimeter law in a system of random surfaces*, Comm. Math. Phys. **92** (1983), pp. 19–69.

[AKN] Aizenman, M., Kesten, H. and Newman, C. M., *Uniqueness of the infinite cluster and continuity of connectivity functions for short and long range percolation*, Comm. Math. Phys. **111** (1987), pp. 505–532.

[AN1] Aizenman, M. and Newman, C. M., *Tree graph inequalities and critical behavior in percolation models*, J. Stat. Phys. **36** (1984), pp. 107–143.

[AN2] Aizenman, M. and Newman, C. M., *Discontinuity of the percolation density in one dimensional* $1/|x-y|^2$ *percolation models*, Comm. Math. Phys. **107** (1986), pp. 611–647.

[BP] Barlow, R. N. and Proschan, F., "Mathematical Theory of Reliability," Wiley, New York, 1965, p. 207.

[Ba] Barsky, D. J., *Critical points and critical exponents in percolation and Ising-type models*, Ph. D. dissertation, Rutgers University (1987).

[BaA] Barsky, D. J. and Aizenman, M., *Percolation critical exponents under the triangle condition*, Ann. Probab. (1991) (to appear).

[BaGN] Barsky, D. J., Grimmett, G. R. and Newman, C. M., *Percolation in half-spaces: equality of critical densities and continuity of the percolation probability*, Prob. Theory and Rel. Fields (to appear).

[BeK] van den Berg, J. and Keane, M., *On the continuity of the percolation probability function.*, "Conference on Modern Analysis and Probability," ed. R. Beals et al., Contemp. Math., 1984, pp. 61–65.

[BezG] Bezuidenhout, C. E. and Grimmett, G. R., *The critical contact process dies out*, Ann. Probab. **18** (1990), pp. 1462–1482.

[BrDS] Bramson, M., Durrett, R. and Swindle, G., *Statistical mechanics of crabgrass*, Ann. Probab. **17** (1989), pp. 444–481.

[BrH] Broadbent, S. R. and Hammersley, J. M., *Percolation processes. I. Crystals and Mazes*, Proc. Camb. Phil. Soc. **53** (1957), pp. 629–641.

[BuK] Burton, R. and Keane, M., *Density and uniqueness in percolation*, Comm. Math. Phys. **121** (1989), pp. 501–505.

[CC1] Chayes, J. T. and Chayes, L., *Critical points and intermediate phases on wedges of* $\mathbb{Z}^d$, J. Phys. A **19** (1986), pp. 3033–3048.

[CC2] Chayes, J. T. and Chayes, L., *On the upper critical dimension of Bernoulli percolation*, Comm. Math. Phys. **113** (1987), pp. 27–48.

[D] Durrett, R., *A new method for proving the existence of a phase transition*, "Spatial Stochastic Processes," eds. K. Alexander and J. Watkins, Birkhäuser, Boston, 1989.

[FKG] Fortuin, C., Kasteleyn, P. and Ginibre, J., *Correlation inequalities on some partially ordered sets.*, Comm. Math. Phys. **22** (1971), pp. 89–103.

[GKN] Gandolfi, A., Keane, M. and Newman, C. M., *Uniqueness of the infinite component in a random graph with applications to percolation and spin glasses*, Submitted to Prob. Theory and Rel. Fields (1989).

[Gr] Grimmett, G. R., "Percolation," Springer–Verlag, New York, 1989.

[GrM] Grimmett, G. R. and Marstrand, J., *The supercritical phase of percolation is well behaved.*, Proc. Roy. Soc. London, Ser. A (1991) (to appear).

[H1] Hammersley, J. M., *Percolation processes. Lower bounds for the critical probability*, Ann. Math. Statist. **28** (1957), pp. 790–795.

[H2] Hammersley, J. M., *Bornes supérieures de la probabilité critique dans un processus de filtration*, "Le Calcul des Probabilités et ses Applications," CNRS, Paris, 1959, pp. 17–37.

[HS1] Hara, T. and Slade, G., *The mean-field critical behaviour of percolation in high dimensions*, "Proceedings of the IXth International Conference on Mathematical Physics," eds. B. Simon, A. Truman and I. M. Davies, Adam Hilger, Bristol, 1989, pp. 450–453.

[HS2] Hara, T. and Slade, G., *Mean field critical phenomena for percolation in high dimensions*, Comm. Math. Phys. **128** (1990), pp. 333–391.

[Ha1] Harris, T. E., *A lower bound for the critical probability in a certain percolation process*, Proc. Camb. Phil. Soc. **56** (1960), pp. 13–20.

[Ha2] Harris, T. E., *Additive set-valued Markov processes and graphical methods*, Ann. Probab. **6** (1978), pp. 355–378.

[K1] Kesten, H., *The critical probability of bond percolation on the square lattice equals* $\frac{1}{2}$, Comm. Math. Phys. **74** (1980), pp. 41–59.

[K2] Kesten, H., *Correlation length and critical probabilities of slabs for percolation*, Manuscript (1988).

[N] Nguyen, B. G., *Gap exponents for percolation processes*, J. Stat. Phys. **49** (1987), pp. 235–243.

[R1] Russo, L., *A note on percolation.*, Z. Wahrsch. Verw. Geb. **43** (1978), pp. 39–48.

[R2] Russo, L., *On the critical percolation probabilities*, Z. Wahrsch. Verw. Geb. **56** (1981), pp. 229–237.

[S] Sarkar, T. K., *Some lower bounds of reliability*, Tech. Report No. 124, Dept. of Operations Research and Statistics, Stanford University, Stanford, CA (1968).

[SW] Seymour, P. D. and Welsh, D. J. A., *Percolation probabilities on the square lattice*, Advances in Graph Theory, ed. B. Bollobás, Ann. Disc. Math. **3** (1978), pp. 277–245.

[T] Tasaki, H., *Hyperscaling inequalities for percolation*, Comm. Math. Phys. **113** (1987), pp. 49–65.


Research supported in part by an NSF Postdoctoral Fellowship (D. J. B.), the University of Arizona Center for the Study of Complex Systems (G. R. G.), NSF Grant No. DMS-8514834 (C. M. N.), and AFOSR Contract No. F49620-86-C0130 to the Arizona Center for Mathematical Sciences under the U. R. I. Program.

David J. Barsky
Department of Mathematics
University of California, Davis
Davis, CA 95616

Geoffrey R. Grimmett
School of Mathematics
University of Bristol
Bristol, England BS8 1TW

Charles M. Newman
Courant Institute of Mathematical Sciences
New York University
251 Mercer St.
New York, NY 10012

# Asymptotics in High Dimensions For the Fortuin-Kasteleyn Random Cluster Model

Harry Kesten

*Abstract.* We consider the Fortuin-Kasteleyn cluster model which is related to the percolation model and the Potts model. We give the asymptotic behavior of the critical probability and the percolation probability as the dimension $d$ tends to infinity, in the case where $Q$ (which corresponds to the number of colors in the Potts model) lies between 1 and 2.

## §1 Introduction

In a paper [Ha], which was far ahead of its time, Harris developed the basic properties of two dimensional percolation. I have personally greatly profited from this paper and am happy to acknowledge this debt in this forum. We now have a fair understanding of two dimensional percolation, and since the recent work of Hara and Slade [HS], also of percolation in high dimensions. In [Ks] we investigated some properties of the critical probability and the percolation probability in the limit when the dimension goes to $\infty$. Most of these properties are shared by the *Fortuin-Kasteleyn* model for $1 < Q < 2$ (see below) and we wish to present here the analogues of [Ks] for this model.

We did not prove Theorem 2 in [Ks] because it is better to prove that theorem for percolation as a special case of the Fortuin-Kasteleyn model corresponding to $Q = 1$. Thus Theorem 2 below includes Theorem $2B$ of [Ks]. We do not prove Theorem $2S$ of [Ks]; the reader should be able to fill this in since its proof is similar to that of Theorem 2 below.

We turn to a description of the model. Fortuin and Kasteleyn [FK] proved that there is a close relationship between the partition function of the Ising model and a certain expectation for bond percolation. This relation has been the heuristic basis for many parallels between the Ising model (or more generally the Potts model) and percolation. Recently it was shown in [ACCN] and [ES] (see also some of their references) that the relationship between the Potts model and bond percolation is much stronger than suggested by the Fortuin-Kasteleyn result. In fact in these references the Potts model and a certain cluster model with a parameter $Q$ (corresponding to the number of colors in the Potts model) can be simultaneously constructed on the same probability space. We shall only describe the part of the construction which we need to formulate our results. We consider configurations of occupied and vacant bonds of $\mathbb{Z}^d$, These configurations will be chosen according to certain probability measures constructed as follows. $\omega$ stands for a generic configuration. For any rectangular box $\Lambda$ in $\mathbb{Z}^d$ consider the probability measure on configurations of occupied and vacant bonds in $\Lambda$ which assigns to the configuration $\omega$ the probability

$$\mu_\Lambda(\omega) = \mu_{p,d,Q,\Lambda}(\omega) = (Z_\Lambda)^{-1} p^{o(\omega)} (1-p)^{v(\omega)} Q^{C(\omega)} \tag{1.1}$$

where

$$\begin{aligned} o(\omega) &= \text{ number of occupied bonds of } \omega \text{ in } \Lambda, \\ v(\omega) &= \text{ number of vacant bonds of } \omega \text{ in } \Lambda, \\ C(\omega) &= \text{ number of components of } \omega \text{ in } \Lambda. \end{aligned}$$

Here a *component (or cluster)* of $\omega$ is a maximally connected set of *occupied* bonds. By some abuse of notation we shall also speak of the vertices in a cluster; these are the endpoints of the bonds in the cluster. We make the convention that a site of $\mathbb{Z}^d$ which has no occupied bonds incident to it is counted as a component by itself. $Z_\Lambda$ is a normalization factor to make (1.1) into a probability measure. Actually (1.1) is still ambiguous, because there are different ways of counting components of a configuration $\omega$ in a box $\Lambda$, corresponding to different *boundary conditions.* We shall only consider the two simplest boundary conditions, namely the *free* and the *wired* boundary conditions. In the free boundary condition we ignore all edges not in $\Lambda$, so that two vertices are in the same component if and only if they are connected by an occupied path in $\Lambda$. In the wired boundary condition all points on the boundary of $\Lambda$ are assumed to be connected to each other; thus with this boundary condition two points belong to the same component if they are connected by an occupied path in $\Lambda$ or if both are connected by occupied paths to the boundary of $\Lambda$. If necessary we shall indicate which boundary condition is being used by adding the superscript $f$ or $w$ (for free and wired, respectively) to $\mu$ and $Z$. The subscript $Q$ will almost always be suppressed.

It is shown in [ACCN] that if $Q > 1$, then the probability measures $\mu_\Lambda^f$ and $\mu_\Lambda^w$ converge weakly to probability measures as $\Lambda$ increases to all of $\mathbb{Z}^d$. The limit measures on the configurations of occupied and vacant bonds of $\mathbb{Z}^d$ will be denoted by $\mu^f$ and $\mu^w$, respectively. Sometimes, when we wish to emphasize the value of $p$ (or $p$ and $d$) we use $\mu_p^f$ and $\mu_p^w$ (respectively $\mu_{p,d}^f$ and $\mu_{p,d}^w$). We call the models governed by $\mu^f$ or $\mu^w$ *Fortuin-Kasteleyn* (or FK for short) random cluster models. When $Q = 1$ the bonds are independent and both $\mu^f$ and $\mu^w$ give us simply the Bernoulli bond percolation model. When $Q$ is an integer greater than 1, then the model corresponds to a Potts model with $Q$ colors. In fact if one gives each of the clusters in the Fortuin-Kasteleyn model (corresponding to $Q$) independently of all other clusters one of the colors $1, \dots, Q$, with all the colors equally likely, then the resulting color distribution (which we now view as a color distribution on the *sites* of $\mathbb{Z}^d$) is the color distribution of a Potts model (cf.[ACCN] and [ES]). In particular for $Q = 2$ the resulting model is equivalent with the Ising model; the two colors of the site $x$ now correspond to the states +1 and -1 of the spin $\sigma_x$ at $x$, and the probability of the configuration is (formally) proportional to

$$\exp\left(\frac{\beta}{2}\Sigma\sigma_x\sigma_y\right), \tag{1.2}$$

where the sum runs over nearest neighbor pairs $\{x, y\}$ and the relation between $p$ and $\beta$ is given by

$$p = 1 - e^{-\beta} \tag{1.3}$$

We shall discuss the Potts model further in another paper with R.Schonmann [KS]. However, $Q$ does not have to be an integer and here we restrict ourselves to the $FK$ model with $1 \leq Q \leq 2$. We need a few more definitions to state our results.

$$C(\mathbf{0}) = \text{ the cluster of the origin in the FK model}$$

(this cluster consists of $\mathbf{0}$ only if all edges incident to the origin are vacant),

$$|A| = \text{ number of vertices in the set } A.$$

For $* = f$ or $w$

$$\theta^*(p, Q) = \theta^*(p, \mathbb{Z}^d, Q) = \mu^*\{C(\mathbf{0}) \text{ is infinite}\}$$
$$p_c^*(\mathbb{Z}^d, Q) = \sup\{p : \theta^*(p, \mathbb{Z}^d, Q) = 0\}.$$

$\theta^*$ and $p_c$ are the *percolation and critical probability*, respectively. The first result of this paper gives the asymptotic behavior of $p_c$.

**Theorem 1.** *In the FK model with $1 \le Q \le 2$ one has*

$$\frac{Q}{2d} + O\left(d^{-2}\right) \le p_c^w \le p_c^f \le \frac{Q}{2d} + O\left[\frac{(\log\log d)^2}{d\log d}\right]. \tag{1.4}$$

The second theorem deals with the limit of the percolation probability when we take $p = \gamma Q/(2d)$ for some fixed $\gamma > 1$ and let $d$ tend to $\infty$. It also shows that vertices behave in this limit independently with regard to the property of belonging to an infinite cluster. We should note that Theorem 2 expresses the generally held belief that in high dimensions the system exhibits "mean field behavior". Certain aspects of this belief for percolation and other models have been confirmed by the work in [HS] and some of its references. For the FK model this means that around a fixed site $x$ the number of neighbors of $x$ connected to $\infty$ is close to its expected value $2d\theta(p)$, irrespective of the state of the edges incident to $x$ itself. Once this is accepted it is easy to derive a consistency relation for $\theta(p)$. This is the so called mean field equation; at $p = Q\gamma/(2d)$ the limit of this equation as $d \to \infty$ is just (1.6).

**Theorem 2.** *In the FK model with $1 \le Q \le 2$ and $\gamma > 1$ fixed*

$$\lim_{d\to\infty} \theta^*\left(\frac{Q\gamma}{2d}, \mathbb{Z}^d, Q\right) = y(\gamma), \tag{1.5}$$

*where $y(\gamma)$ is the unique solution of*

$$y = \frac{1-\exp(-Q\gamma y)}{1+(Q-1)\exp(-Q\gamma y)}. \tag{1.6}$$

*More generally, for fixed $\gamma > 1$*

$$\begin{aligned}\lim_{d\to\infty}\sup_{A,B}\Big|\mu^*_{\frac{Q\gamma}{2d}}\{ & \text{ all sites in } A \text{ belong to an infinite} \\ & \text{occupied cluster, but none of the sites in } B \text{ do}\} \\ & - (y(\gamma))^{|A|}(1-y(\gamma))^{|B|}\Big| = 0.\end{aligned} \tag{1.7}$$

*The sup in (1.7) is over all pairs of finite disjoint sets $A$ and $B$. Finally, if $\gamma < 1$, then*

$$\theta^*\left(\frac{Q\gamma}{2d}, \mathbb{Z}^d, Q\right) = 0 \text{ for all sufficiently large } d. \tag{1.8}$$

*Remark 1.* (1.5) - (1.7) remain valid for $Q > 2$, even for a $\gamma \leq 1$, as long as $\theta^*(Q\gamma(2d)^{-1}, \mathbb{Z}^d, Q)$ is bounded away from zero for large $d$. The difficulty is that we do not know that the latter condition holds for all $\gamma$ with $Q\gamma > \limsup_{d\to\infty} 2dp_c(\mathbb{Z}^d, Q)$ (see Remark 5 at end and [KS, Theorem 1]). For integral $Q > 2$ and the wired boundary condition, the analogue of Theorem 2 follows everywhere above criticality from [KS, Theorem 1], since in such a case the percolation probability equals the magnetization of the $Q$ color Potts model (see [ACCN, Lemma 2.1]).

*Remark 2.* For $Q = 1$ it was known that with probability 1 there is at most one infinite occupied cluster (see [AKN] or [GGR]); for general $Q > 1$ this was only recently proved in [BK] (see also [GNK]).

*Remark 3.* Results very similar to Theorem 2 for the contact process have appeared in [SV]; presumably there is also an analogue for directed percolation. The analogue of Theorem 2 for integral $Q \geq 2$ (or equivalently, the Potts model with $Q$ colors) can be found in [KS].

*Acknowledgement.* The author is indebted to R. Schonmann for many helpful suggestions for this paper.

## §2 Proof of Theorem 1.

We write $E^*_\Lambda$ and $E^*$ for expectation with respect to $\mu^*_\Lambda$ and $\mu^*$, respectively. For any configuration $\omega$ of occupied and vacant edges we set, as usual,

$$\omega(e) = \begin{cases} 1 & \text{if } e \text{ is occupied,} \\ 0 & \text{if } e \text{ is vacant.} \end{cases}$$

$\omega_1 \prec \omega_2$ means $\omega_1(e) \leq \omega_2(e)$ for all $e$. $\omega_1 \wedge \omega_2$ is the configuration with $(\omega_1 \wedge \omega_2)(e) = \omega_1(e) \wedge \omega_2(e)$ for all $e$, and similarly for $\omega_1 \vee \omega_2$ ($a \wedge b$ denotes $\min(a,b)$ and $a \vee b$ denotes $\max(a,b)$). We identify the space of configurations with

$$\Omega = \Omega_d := \{0,1\}^{\mathcal{E}_d}$$

where $\mathcal{E}_d$ is the edge set of $\mathbb{Z}^d$. A function f on the space of configurations is called *increasing (decreasing)* if $f(\omega_1) \leq f(\omega_2)$ whenever $\omega_1 \prec \omega_2$ (respectively, $\omega_2 \prec \omega_1$). For two measures $\mu_1, \mu_2$ on $\Omega_d$ we say that $\mu_1 \leq \mu_2$ if

$$\int f d\mu_1 \leq \int f d\mu_2 \text{ for all increasing, positive cylinder functions } f.$$

($f$ is a cylinder function if it depends on finitely many $\omega(e_i)$ only.) We further define for any *finite* set of edges $E$ and configuration $\eta$,

$$\eta_E = \text{ restriction of } \eta \text{ to } E,$$

$$\mu^*\{\cdot|\eta_E\} = \text{ conditional measure in the } FK \text{ model with boundary condition } *, \text{ given } \omega(e) = \eta(e) \text{ for } e \in E.$$

For an arbitrary set of edges $E$ we set

$$\mathcal{F}_E = \sigma - \text{field generated by } \{\omega(e) : e \in E\}.$$

We note that for any finite set of edges $E$, and fixed configuration $\eta$

$$\text{(2.1)} \quad \mu^*\{\omega(e) = \eta(e), e \in E|\mathcal{F}_{E^c}\} \geq [\frac{p}{Q} + (1-p)]^{-o(\eta_E)} \left(\frac{p}{Q}\right)^{o(\eta_E)} (1-p)^{v(\eta_E)} > 0,$$

where $o(\eta_E)$ and $v(\eta_E)$ are the number of occupied and vacant edges of $\eta$ in $E$, respectively. (2.1) follows directly from (1.1). Indeed the probability of any edge $e$ to be occupied given all other edges is at least $pQ^{-1}[p/Q + (1-p)]^{-1}$, as shown in [ACCN, formula (2.10)]. Similarly, the conditional probability to be vacant is at least $(1-p)$. This last estimate shows that also

$$\text{(2.2)} \qquad \mu^*\{\omega(e) = 1 \text{ for all } e \in E|\mathcal{F}_{E^c}\} \leq p^{|E|}$$

Of course, for finite $E, \mu^*\{\cdot|\eta_E\}$ is simply defined by

$$\mu^*\{A|\eta_E\} = [\mu^*\{\omega(e) = \eta(e), e \in E\}]^{-1} \cdot \mu^*\{A \text{ and } \omega(e) = \eta(e), e \in E\}$$

This formula is meaningful, since the denominator is not zero by (2.1).

We use the strong $FKG$ property for the $FK$ measures in the following form.

**Lemma 1.** *If $\eta' \prec \eta''$ (or even $\eta'_E \prec \eta''_E$) and $E$ is finite, then for $* = f$ or $w$*

$$\mu^* \{\cdot|\eta'_E\} \leq \mu^* \{\cdot|\eta''_E\} \tag{2.3}$$

*(In other words, for increasing cylinder functions $f, E^*\{f|\omega = \eta \text{ on } E\}$ is increasing in $\eta$).*

*Proof.* It is verified in [ACCN] that $\mu^*_\Lambda$ satisfies the lattice condition

$$\mu^*_\Lambda \{\omega_\Lambda = \gamma'_\Lambda \wedge \gamma''_\Lambda\} \mu^*_\Lambda \{\omega_\Lambda = \gamma'_\Lambda \vee \gamma''_\Lambda\} \geq \mu^*_\Lambda \{\omega_\Lambda = \gamma''_\Lambda\} \mu^*_\Lambda \{\omega_\Lambda = \gamma'_\Lambda\} \tag{2.4}$$

for any given configurations $\gamma'$ and $\gamma''$. Applying (2.4) to configurations $\gamma'$ which equal $\eta'$ on $E$ and configurations $\gamma''$ which equal $\eta''$ on $E$, with $\eta' \prec \eta''$, we see that for $E \subset \Lambda$ one also has

$$\begin{aligned} &\mu^*_\Lambda \{\omega_\Lambda = \gamma'_\Lambda \vee \gamma''_\Lambda |\, \eta''_E\} \mu^*_\Lambda \{\omega_\Lambda = \gamma'_\Lambda \wedge \gamma''_\Lambda | \eta'_E\} \\ &\qquad \geq \mu^*_\Lambda \{\omega_\Lambda = \gamma''_\Lambda | \eta''_E\} \mu^*_\Lambda \{\omega_\Lambda = \gamma'_\Lambda | \eta'_E\} \end{aligned} \tag{2.5}$$

It is well known that (2.5) implies (2.3) with $\mu$ replaced by $\mu_\Lambda$ (c.f. [Ho],[Km]). Alternatively, one can apply Theorem 2.2 of [BB]. Finally, one obtains (2.3) by letting $\Lambda$ increase to $\mathbb{Z}^d$ (cf. Theorem 2.3 in [ACCN]).

**Corollary 2.** *Let $E_1 \subset E_2 \subset \dots$ be an increasing sequence of finite sets of edges of $\mathbb{Z}^d$, and let $E_\infty = \cup E_i$. Define*

$$\eta^0 = \text{ configuration with } \eta^0(e) \text{ equal to } 0 \text{ for all } e.$$

*Then for any $\sigma$field $\mathcal{G} \subset \mathcal{F}_{E_\infty}$*

$$\mu^* \{\cdot|\mathcal{G}\} \geq \lim_{n\to\infty} \mu^* \{\cdot|\eta^0_{E_n}\} \ a.e.\ [\mu^*]$$

*in the sense that for any bounded increasing cylinder function $f$*

$$E^* \{f|\mathcal{G}\} \geq \lim_{n\to\infty} E^* \{f|\eta^0_{E_n}\} \ a.e.\ [\mu^*]. \tag{2.6}$$

*(The existence of the limits in the right hand side of (2.6) is part of the statement of the corollary.)*

*Proof.* Fix a bounded increasing cylinder function $f$. For $n < m$, $E_n \subset E_m$, and hence

$$E^*\{f|\eta^0_{E_n}\} = \sum_{\bar\eta} \mu^*\{\omega_{E_m \backslash E_n} = \bar\eta|\eta^0_{E_n}\}E^*\{f|\omega_{E_n} = \eta^0_{E_n}, \omega_{E_m\backslash E_n} = \bar\eta\}$$

$$\geq E^*\{f|\omega_{E_m} = \eta^0_{E_m}\}. \tag{2.7}$$

($\bar\eta$ here runs over all possible configurations on $E_m \backslash E_n$). Thus the sequence of numbers $E^*\{f|\eta^0_{E_n}\}$ is decreasing in $n$, and the limit in the right hand side of (2.6) exists.

Essentially the same argument as in (2.7) shows that

$$E^*\{f|\mathcal{F}_{E_n}\} \geq E^*\{f|\eta^0_{E_n}\}.$$

By taking the limit $n \to \infty$ and applying the martingale convergence theorem we obtain (2.6) when $\mathcal{G} = \mathcal{F}_{E_\infty}$. Finally, for $\mathcal{G} \subset \mathcal{F}_{E_\infty}$ (2.6) follows from

$$E^*\{f|\mathcal{G}\} = E^*\{E^*\{f|\mathcal{F}_{E_\infty}\}|\mathcal{G}\}$$
$$\geq E^*\left\{\lim_{n\to\infty} E^*\{f|\eta^0_{E_n}\}|\mathcal{G}\right\} = \lim_{n\to\infty} E^*\{f|\eta^0_{E_n}\}$$

(recall that the last member is a constant).

**Corollary 3.** $\mu^w\{\cdot\} \geq \mu^f\{\cdot\}$.

*Proof.* Let

$$\Lambda_n := \{x : |x| \leq n\}, \tag{2.8}$$

where $|x| = \sum_1^d |x_i|$ for $x = (x_1, \dots, x_d)$ (the $\ell^1$ norm). One easily sees that for any bounded function $f$ which depends only on the edges in $\Lambda_n$ one has

$$\mu^w_{\Lambda_n}\{f\} = \mu^f\{f|\omega(e) = 1 \text{ for all } e \text{ in } \Lambda_{n+1}\backslash(\text{interior of } \Lambda_n)\}.$$

Also

$$\mu^f_{\Lambda_n}\{f\} = \mu^f\{f|\omega(e) = 0 \text{ for all } e \text{ in } \Lambda_{n+1}\backslash\Lambda_n\}.$$

Again an argument similar to (2.7) shows that if $f$ is in addition increasing, then

$$\mu^w_{\Lambda_n}\{f\} \geq \mu^f_{\Lambda_n}\{f\}$$

The corollary follows by letting $n \to \infty$.

*Proof of Theorem 1.* (1.4) is a simple combination of the comparison theorems of [ACCN] and the known asymptotic behavior of $p_c(\mathbb{Z}^d, Q)$ for $Q = 1, 2$. Indeed, for $Q = 1$, the case of percolation, the boundary condition is irrelevant, and by Theorem 1 of [Ks]

$$\frac{1}{2d-1} \leq p^f_c\left(\mathbb{Z}^d, 1\right) = p^w_c\left(\mathbb{Z}^d, 1\right) \leq \frac{1}{2d} + O\left[\frac{(\log\log d)^2}{d\log d}\right]. \tag{2.9}$$

In the remainder of this proof $\beta$ and $p$ are given in terms of each other by

$$p = 1 - e^{-\beta}.$$

Correspondingly, we define $\beta^*_c$ by

$$p^*_c\left(\mathbb{Z}^d, Q\right) = 1 - \exp\left(-\beta^*_c\left(\mathbb{Z}^d, Q\right)\right), * = f \text{ or } w. \tag{2.10}$$

First we note that Cor. 3 implies that for any $p, Q$

$$\theta^f\left(p, \mathbb{Z}^d, Q\right) \leq \theta^w\left(p, \mathbb{Z}^d, Q\right), \tag{2.11}$$

and consequently

$$p^f_c\left(\mathbb{Z}^d, Q\right) \geq p^w_c\left(\mathbb{Z}^d, Q\right). \tag{2.12}$$

Next we note that the comparison principle of Fortuin [Fo, Lemma 3], which is also treated in [ACCN, Theorem 4.1], shows that

$$\frac{\beta'}{Q'} \geq \frac{\beta}{Q} \text{ and } Q' \geq Q \geq 1 \tag{2.13}$$

implies that (with $p' = 1 - \exp(-\beta')$)

$$\theta^*\left(p', \mathbb{Z}^d, Q'\right) \geq \theta^*\left(p, \mathbb{Z}^d, Q\right) \tag{2.14}$$

We first apply this with $Q = 1$. We then obtain

$$\theta^*\left(p', \mathbb{Z}^d, Q'\right) > 0 \text{ as soon as } \beta' > Q'\beta_c^*\left(\mathbb{Z}^d, 1\right).$$

Combined with (2.9) this proves
(2.15)

$$p_c^*\left(\mathbb{Z}^d, Q'\right) \leq 1 - \exp\left(-Q'\beta_c^*\left(\mathbb{Z}^d, 1\right)\right) \leq \frac{Q'}{2d} + O\left[\frac{(\log\log d)^2}{d\log d}\right], Q' \geq 1.$$

When we apply (2.14) with $Q' = 2$ we obtain in the same way

$$\theta^*\left(p, \mathbb{Z}^d, Q\right) = 0 \text{ for } \beta < \frac{Q}{2}\beta_c^*\left(\mathbb{Z}^d, 2\right). \tag{2.16}$$

Now we use (cf. [ACCN,Theorem 2.3(iii) or Lemma 2.1]) that percolation occurs in the FK model with $Q = 2$ and with the wired boundary condition, if and only if the spontaneous magnetization in the corresponding Potts model with $Q = 2$ does not vanish. This corresponding Potts model is precisely the Ising model with probability measure described by (1.2). It is known that the magnetization in this Ising model is zero for $\tanh(\beta/2) < (2d-1)^{-1}$ (cf. [Fi]). Thus

$$\tanh\left(\beta_c^w\left(\mathbb{Z}^d, 2\right)/2\right) \geq \frac{1}{2d-1}. \tag{2.17}$$

Together with (2.16) this shows

$$\begin{aligned} p_c^w\left(\mathbb{Z}^d, Q\right) &\geq 1 - \exp\left(-\frac{Q}{2}\beta_c^w\left(\mathbb{Z}^d, 2\right)\right) \\ &\geq \frac{Q}{2d} + O\left(d^{-2}\right), \quad d \to \infty, \quad 1 \leq Q \leq 2. \end{aligned} \tag{2.18}$$

(1.4) follows from (2.12), (2.15) and (2.18).

## §3 Proof of Theorem 2.

We use $A \leftrightarrow B$ to denote the event that there exists an occupied path from some vertex in $A$ to some vertex in $B$. $A \leftrightarrow B$ in $C$ occurs if this path can be chosen to lie in $C$. $A \not\leftrightarrow B$ and $A \not\leftrightarrow B$ in $C$ are the complements of the events $A \leftrightarrow B$ and $A \leftrightarrow B$ in $C$, respectively.

The basic idea of the proof of Theorem 2 is the same as for the theorem of [BKLS]. First it is shown that $\theta$ satisfies a consistency relation (cf. Lemma 4 below). This relation merely expresses a kind of Markov property. The right hand side of this relation involves the random variable $M/(2d)$, where

(3.1)

$$I(x) = \text{indicator function of the event } \{x \text{ is connected to } \infty \text{ by an occupied path which does not use any edge incident to } \mathbf{0}\},$$

and

$$M = \sum_{\substack{x \text{ adjacent} \\ \text{to } \mathbf{0}}} I(x). \tag{3.2}$$

In order to relate the right hand side of (3.4) to $\theta$ we show that

$$\frac{1}{2d} M - E^* \{I(\xi_1)\} \to 0 \text{ in distribution as } d \to \infty \tag{3.3}$$

($\xi_1 = (1, 0, \dots, 0)$). This is done by showing that the covariance at $I(x)$ and $I(y)$ tends to 0. We then also have to show that $E^*\{I(e_1)\} - \theta^* \to 0$, but the main difficulty is the covariance estimate. Once this is done the argument is identical to that of [BKLS].

To show that the covariance of $I(x)$ and $I(y)$ is small for large $d$ (cf. Lemmas 5-8), we prove that $I(x)$ may be replaced by the indicator function of $\{x \leftrightarrow \text{boundary of } \Lambda_n\}$ for some large $n$, independent of $d$. We also prove that the events $\{x \leftrightarrow \text{boundary of } \Lambda_n\}$ and $\{y \leftrightarrow \text{boundary of } \Lambda_n\}$ are approximately independent for large $d$; the intuitive reason for this is that the clusters of $x$ and $y$ in $\Lambda_n$ are, with high probabiiity, disjoint when $d$ is large. However, as we shall see in the proof of Theorem 2, the details are more complicated. In fact a direct proof of the asymptotic negligability of the covariance of $I(x)$ and $I(y)$ is available only when $p = p(d) = \gamma(2d)^{-1}$ for *all but countably many* fixed $\gamma > 1$, and when $d \to \infty$ along suitable subsequences.

**Lemma 4.** *With* $p = 1 - e^{-\beta}$

$$\theta^*\left(p, Z^d, Q\right) = E^*\left\{\frac{1 - \exp(-\beta M)}{1 + (Q-1)\exp(-\beta M)}\right\}. \tag{3.4}$$

*Proof.* This lemma is obtained by calculating $\theta^* = \mu^*\{\mathbf{0} \leftrightarrow \infty\}$ by first conditioning on all the edges not incident to $\mathbf{0}$. As explained in [KS, Sect. 4] it is easy to see that the conditional probability of $\mathbf{0} \leftrightarrow \infty$ given all these edges is

$$\frac{\left[1 - (1-p)^M\right] Q^{-1}}{\left[1 - (1-p)^M\right] Q^{-1} + (1-p)^M} = \frac{1 - \exp(-\beta M)}{1 + (Q-1)\exp(-\beta M)}.$$

(3.4) is immediate from this; if necessary, the reader can find more details in [KS].

Let $\Lambda_n$ be as in (2.8) and $\partial\Lambda_n$ be its boundary, i.e. $\partial\Lambda_n = \{x : |x| = n\}$. (Recall that $|x|$ denotes the $\ell^1$ norm.) Set

$$\Gamma_n(x) = \{y \in \partial\Lambda_n : x \leftrightarrow y \text{ in } \Lambda_n\} \text{ and } \nu_n(x) = |\Gamma_n(x)|.$$

**Lemma 5.** *Let $\gamma$ and $\ell$ be fixed. Then there exists a $0 < C_1 = C_1(\gamma, \ell) < \infty$ such that for each $x$ and $n \ge |x|$ and each $d$*

$$\begin{aligned}\sup_{p \le \frac{1}{2}\wedge\gamma(2d)^{-1}} \mu^*_{p,d}\{&x \leftrightarrow \partial\Lambda_n \textit{ but } \nu_m(x) \le \ell \textit{ for more than} \\ &j \textit{ values of } |x| \le m \le n\} \\ &\le (1 - C_1)^j, \quad j = 1, 2, \ldots\end{aligned} \tag{3.5}$$

*Proof.* Write $\mathcal{F}_m$ for $\mathcal{F}_{\Lambda_m}$, the $\sigma$-field generated by all edges in $\Lambda_m$. Then $\nu_m(x)$, as well as the set $\Gamma_m(x)$, are $\mathcal{F}_m$-measurable. Moreover, it follows from (2.1) that for $p \le 1/2$

$$\begin{aligned}&\mu^*_{p,d}\{x \leftrightarrow \partial\Lambda_m \text{ but } x \not\leftrightarrow \partial\Lambda_{m+1} | \mathcal{F}_m\} \\ &\quad = \mu^*_{p,d}\{\text{all edges not in } \Lambda_m \text{ but incident} \\ &\qquad\qquad \text{to a } y \in \Gamma_m(x) \text{ are vacant } |\mathcal{F}_m\} \\ &\quad \ge (1-p)^{2d\nu_m(x)} \\ &\quad \ge \exp(-4dp\nu_m(x)).\end{aligned}$$

If in addition $p \le \gamma(2d)^{-1}$, then the last expression is at least $\exp(-2\gamma\ell)$ on the event $\{\nu_m(x) \le \ell\}$. Now define

$$m_1 = |x|, m_{i+1} = \inf\{m > m_i : \nu_m(x) \le \ell\} \tag{3.6}$$

($= \infty$ if no such $m$ exists). The $m_i$ are the successive values of $m$ for which $\nu_m(x) \le \ell$. By the above estimates we have for $n \ge |x|$

$$\begin{aligned}
&\mu^*_{p,d}\{x \leftrightarrow \partial\Lambda_n \text{ and } m_1 < m_2 < \ldots < m_j < n\} \\
&\quad \le \mu^*_{p,d}\{m_1 < \ldots < m_j < \infty \text{ and for each } i \le j \text{ at least one edge} \\
&\qquad\qquad \text{outside } \Lambda_{m_i} \text{ and incident to } \Gamma_{m_i}(x) \text{ is occupied}\} \\
&\quad \le E^*\{E^*\{ \text{ there exists at least one occupied edge not in } \Lambda_{m_j} \text{ and} \\
&\qquad \text{incident to } \Gamma_{m_j}|\mathcal{F}_{m_j}\}I[m_1 < \ldots < m_j < \infty \text{ and for each } i \le j-1 \\
&\qquad \text{at least one edge outside } \Lambda_{m_i} \text{ and incident to } \Gamma_{m_i}(x) \text{ is occupied}]\} \\
&\quad \le [1 - \exp(-2\gamma\ell)]\mu^*_{p,d}\{m_1 < \ldots < m_{j-1} < \infty \text{ and for each } i \le j-1 \text{ at} \\
&\qquad \text{least one edge outside } \Lambda_{m_i} \text{ and incident to } \Gamma_{m_i}(x) \text{ is occupied}]\} \\
&\quad \le \ldots \le [1 - \exp(-2\gamma\ell)]^j.
\end{aligned}$$

In addition to $\Gamma_m(x)$ we need the closely related entities

$$\begin{aligned}
\tilde{\Delta}_m(x) = \{y \in \Lambda_m : x \leftrightarrow y \quad &\text{via an occupied path with succesive vertices} \\
&x = y_0, y_1, \ldots, y_\lambda = y \quad |y_i| = |x| + i\},
\end{aligned}$$

and

$$\Delta_m(x) = \{y \in \Lambda_m : x \leftrightarrow y \in \Lambda_m\}.$$

We want to show that $\Delta_m$ is not too large, and in fact equals $\tilde{\Delta}_m$ with high probability.

**Lemma 6.** *Let $\xi$ be a fixed unit vector. Then*

$$E^*_{p,d}\left\{\left|\tilde{\Delta}_m(x)\right|\right\} \le \sum_{\lambda=0}^{m} (p \cdot 2d)^\lambda, \tag{3.7}$$

$$\mu^*_{p,d}\{\tilde{\Delta}_m(x) \neq \Delta_m(x), \text{ or } \Delta_m(x) \text{ contains a vertex } z \text{ such that the edge from } z \text{ to } z+\xi \text{ is occupied}\} \leq p\,(m+1) \sum_{\lambda=0}^{m} (p \cdot 2d)^{\lambda}. \tag{3.8}$$

*Proof.* (3.7) is easy, since the number of paths with vertices $x = y_0, y_1, \dots, y_\lambda$ with $|y_i| = |x| + i, 0 \leq i \leq \lambda$, is at most $(2d)^\lambda$. These paths are necessarily selfavoiding. By (2.2), the probability that a given path of this kind has all its edges occupied is at most $p^\lambda$. If the path is to lie in $\Lambda_m$, then $\lambda$ can be at most $m$. This implies (3.7).

We turn to (3.8). Assume that $\Delta_m(x) \neq \tilde{\Delta}_m(x)$. Then there exists a selfavoiding path all of whose edges are occupied, and with successive vertices $y_0 = x, y_1, \dots, y_\lambda$ in $\Lambda_m$, but such that $|y_i| < |x| + i$ for some $i$. Let $i$ be the minimal index with this property. Then $i \leq m+1$, each step $y_j - y_{j+1}, 1 \leq j \leq i$, is of the form $\pm\xi_{r_j}$ (where $\xi_r$ is the $r$-th coordinate vector), and $y_i - y_{i-1} = -(y_j - y_{j-1})$ for some $j < i$. Therefore,

$$\begin{aligned}
&\mu^*_{p,d}\{\Delta_m(x) \neq \tilde{\Delta}_m(x)\} \\
&\quad \leq \mu^*_{p,d}\{\exists \text{ selfavoiding path of } i \leq m+1 \text{ occupied edges and successive vertices } y_0 = x, y_1, \dots, y_i \text{ such that } y_i - y_{i-1} = -(y_j - y_{j-1}) \text{ for some } 1 \leq j < i\} \\
&\quad \leq \sum_{i=1}^{m+1} (i-1)\,(2d)^{i-1} p^i \leq mp \sum_{\lambda=0}^{m} (p \cdot 2d)^{\lambda}
\end{aligned}$$

(The factor $(i-1)$ in the one but last member represents the maximal number of choices for $y_i - y_{i-1}$ once $y_0, \dots, y_{i-1}$ are given.) To obtain (3.8) we merely have to add the probabilities

$$\mu^*_{p,d}\{\exists \text{ self avoiding path of } i \leq m+1 \text{ occupied edges and successive vertices } x = y_0, \dots, y_i \text{ such that } y_i = y_{i-1} + \xi\}$$

$$\leq \sum_{i=1}^{m+1} (2d)^{i-1} p^i.$$

The next lemma basically shows that once $x$ is connected to $\partial\Lambda_n$, for large $n$, there is a high conditional probability that $x$ is actually connected to $\infty$; moreover these estimates are uniform in $d$, as long as $p \geq Q\lambda'(2d)^{-1}$ for some fixed $\lambda' > 1$. Unfortunately the precise statement which we need is more technical.

**Lemma 7.** *Define*

$$B(x,n) = \{x \leftrightarrow \partial\Lambda_{2n}, \text{ but there does not exist an } s \in (n,2n] \text{ and a } y \in \partial\Lambda_s \text{ such that } x \leftrightarrow y \text{ in } \Lambda_s \text{ and } y \leftrightarrow \infty \text{ by an occupied path with all its edges outside } \Lambda_s\}.$$

*Then for each fixed* $k, \gamma'' \geq \gamma' > 1$

(3.9)
$$\lim_{n\to\infty} \limsup_{d\to\infty} \sup \left\{\mu^*_{p,d}\{B(x,n)\} : |x| \leq k, Q\gamma'(2d)^{-1} \leq p \leq \gamma''(2d)^{-1}\right\} = 0.$$

*Proof.* We need the following result from [Ks, Lemma 7]:

$$\liminf_{d\to\infty} \theta^*\left(\frac{\tilde\gamma}{2d}, \mathbb{Z}^d, 1\right) \geq K_5\left[(\tilde\gamma - 1)\wedge 1\right]$$

for some constant $K_5 > 0$ and all $\tilde\gamma$. (Recall that the boundary condition is irrelevant when $Q = 1$.) In particular, by virtue of (2.11), (2.13), (2.14), there exists a $C_2 = C_2(\gamma') > 0$ and $d_0 = d_0(\gamma')$ such that

$$\theta^w(p, Z^d, Q) \geq \theta^f(p, Z^d, Q) \geq C_2 \text{ uniformly for } d \geq d_0, p \geq Q\gamma'(2d)^{-1}. \tag{3.10}$$

Now let $\nu_m$ be as in Lemma 5 and define

$$\sigma = \sigma(n,\ell) = \min\{s : s > n, \nu_s \geq \ell\}$$

Then

$$\mu^*_{p,d}\{B(x,n)\} \leq \mu^*_{p,d}\{x \leftrightarrow \partial\Lambda_{2n} \text{ but } \sigma > 2n\} + \mu^*_{p,d}\{\sigma \leq 2n \text{ but each } y \in \partial\Lambda_\sigma \text{ with } x \leftrightarrow y \text{ in } \Lambda_\sigma \text{ is not connected to } \infty \text{ outside } \Lambda_\sigma\}. \tag{3.11}$$

The first term in the right hand side is equal to

$$\mu^*_{p,d}\{x \leftrightarrow \partial\Lambda_{2n} \text{ but } \nu_s(x) < \ell \text{ for } n < s \le 2n\}$$

By virtue of Lemma 5, this is for $n \ge k \ge |x|$ and $\gamma''(2d)^{-1} \le 1/2$, at most

$$(1 - C_1(\gamma'', \ell))^n. \tag{3.12}$$

To estimate the second term observe that the event $\{\sigma = s\}$ as well as the set $\Gamma_s(x)$ are $\sigma_s$ measurable. It therefore suffices to estimate for fixed $s \in (n, 2n]$, and a fixed set $\Gamma_s = \{y_1, \dots, y_\nu\}$ of vertices in $\partial\Lambda_s$,

(3.13)

$$\mu^*_{p,d}\{\text{ none of the } y_i, 1 \le i \le \nu, \text{ is connected to } \infty \text{ outside } \Lambda_s | \mathcal{F}_{\Lambda_s}\}.$$

The second term in the right hand side of (3.11) is bounded by the maximum of (3.13) over all $s \in (n, 2n]$ and all $\Gamma \subset \partial\Lambda_s$ with $\nu \ge \ell$. To estimate (3.13) we express each $y_i$ as

$$y_i = \pm\xi_{j_1(i)} \pm \xi_{j_2(i)} \pm \cdots \pm \xi_{j_s(i)}.$$

The collection $\{j_1(i), \dots, j_s(i)\}$ is determined by $y_i$, and hence

$$\mathcal{H} = \bigcup_{i \le \nu} \{j_1(i), \dots, j_s(i)\}$$

is also determined by $\{y_1, \dots, y_\nu\}$. Now define the affine spaces

$$S_i = y_i + span\{\xi_j : j \notin \mathcal{H}\}.$$

The $S_i$ are pairwise disjoint, since $y_\alpha - y_\beta \in span\{\xi_j : j \in \mathcal{H}\}$ for each $1 \le \alpha, \beta \le \nu$. For similar reasons

$$S_i \cap \Lambda_s = \{y_i\}.$$

This set contains no edges, so that (3.13) is bounded by

$$\mu^*_{p,d}\{\bigcap_{i=1}^{\nu} (y_i \leftrightarrow \infty \text{ in } S_i)^c | \mathcal{F}_{\Lambda_s}\}. \tag{3.14}$$

The proof of (3.9) will be completed by using Cor. 2 to find a bound for (3.14).

Fix $i$ for the moment, and choose

$$E_m = \Lambda_s \cup \ \{\text{edges of } \Lambda_m \text{ outside } S_i\}.$$

Clearly the $E_m$ are increasing in $m$ to

$$E_\infty = \Lambda_s \cup \ \{\text{edges outside } S_i\}.$$

For the sake of argument consider now the wired boundary conditions. Then, for fixed $t$ and $m$ with $t+1<m$

$$\begin{aligned} &\mu^w_{p,d}\{y_i \leftrightarrow \partial\Lambda_t \text{ in } S_i | \eta^0_{E_m}\} \\ &\quad = \lim_{r\to\infty} \mu^w_{p,d}\{y_i \leftrightarrow \partial\Lambda_t \text{ in } S_i | \omega(e) = 0 \text{ for } e \text{ in } E_m, \omega(e) = 1 \\ &\qquad\qquad \text{for } e \in (\partial\Lambda_r)\cap S_i\} \\ (3.15)\quad &\quad \geq \lim_{r\to\infty} \mu^w_{p,d}\{y_i \leftrightarrow \partial\Lambda_t \text{ in } S_i | \omega(e) = 0 \text{ for } e \text{ in } \Lambda_r \text{ but outside } S_i, \\ &\qquad\qquad w(e) = 1 \text{ for } e \in (\partial\Lambda_r)\cap S_i\}. \end{aligned}$$

We claim that this last conditional probability (before the limit over $r$ is taken) equals

$$(3.16)\qquad \mu^w_{p,\bar{d}}\{0 \leftrightarrow \partial\Lambda_{t-|y_i|} | w(e) = 1 \text{ for } e \in \partial\Lambda_{r-|y_i|}\},$$

where

$$\bar{d} = d - \ (\text{cardinality of } \mathcal{H}).$$

To see this we first observe that under the conditions $w(e)=0$ for $e$ in $\Lambda_r$ but outside $S_i$, and $w(e)=1$ for $e \in \partial\Lambda_r \cap S_i$, the components of the configuration inside $\Lambda_r \cap S_i$ are uniquely determined, independent of the configuration outside $S_i$, or outside $(\Lambda_r \cap S_i)$. Indeed, whether two vertices in $\Lambda_r \cap S_i$ are connected by an occupied path does not depend on the latter edges. By (1.1) this means that the conditional distribution in $\Lambda_r \cap S_i$, given $\omega(e)=0$ for $e$ in $\Lambda_r$ but outside $S_i$ and $\omega(e)=1$ for $e \in (\partial\Lambda_r)\cap S_i$, also does not depend on edges outside $S_i$ or outside $\Lambda_r \cap S_i$. In fact this conditional distribution is the same as if only the edges in $S_i$ exist. The equality of (3.16) and the conditional probability in the last member of (3.15) now follows from the fact that $S_i$ is isomorphic to the $\bar{d}$-dimensional subspace $S := span\{e_j :\notin \mathcal{H}\}$, and $(\partial\Lambda_r) \cap S_i$ is isomorphic to $\partial\Lambda_{r-|y_i|} \cap S$ (under the same isomorphism which takes $S_i$ to $S$).

Substituting (3.16) for the conditional probability in the last member of (3.15) we obtain

$$\mu^w_{p,d}\{y_i \leftrightarrow \partial\Lambda_t \text{ in } S_i|\eta^0_{E_m}\} \geq \mu^w_{p,\bar{d}}\{0 \leftrightarrow \partial\Lambda_{t-|y_i|}\}.$$

Next, by Cor. 2

$$\mu^w_{p,d}\{y_i \leftrightarrow \partial\Lambda_t \text{ in } S_i|\mathcal{F}_{E_\infty}\} \geq \lim_{m\to\infty} \mu^w_{p,d}\{y_i \leftrightarrow \partial\Lambda_t \text{ in } S_i|\eta^0_{E_m}\}$$
$$\geq \mu^w_{p,\bar{d}}\{0 \leftrightarrow \partial\Lambda_{t-|y_i|}\}.$$

The same inequality holds when the wired boundary condition is replaced by the free boundary condition (by exactly the same argument).

Finally, we take the limit $t \to \infty$, to obtain

$$\mu^*_{p,d}\{y_i \leftrightarrow \infty \text{ in } S_i|\mathcal{F}_{E_\infty}\} \geq \mu^*_{p,\bar{d}}\{0 \leftrightarrow \infty\} = \theta^*(p, \mathbb{Z}^{\bar{d}}, Q).$$

But $p \geq Q\gamma'(2d)^{-1}$ and

$$\bar{d} = d - \text{ (cardinality of } \mathcal{H}) \geq d - s\nu_s(x)$$

imply

$$p \geq Q\bar{\gamma}\left(2\bar{d}\right)^{-1} \text{ with } \bar{\gamma} = 1 + \tfrac{1}{2}(\gamma' - 1)$$

for large enough $d$, provided $s\nu_s(x) \leq (2\gamma')^{-1}(\gamma' - 1)\,d$. If $p \geq Q\bar{\gamma}\left(2\bar{d}\right)^{-1}$, then we have by (3.10), for large enough $\bar{d}$,

$$\theta^*\left(p, \mathbb{Z}^{\bar{d}}, Q\right) \geq C_2(\bar{\gamma}) > 0.$$

From the above we obtain that on $\left\{s\nu_s(x) \leq (2\gamma')^{-1}(\gamma' - 1)\,d\right\}$

$$\mu^*_{p,d}\{y_i \leftrightarrow \infty \text{ in } S_i|\mathcal{F}_{E_\infty}\} \geq C_2(\bar{\gamma}) \tag{3.17}$$

for large $d$. Since $S_j$ is disjoint from $S_i$ for $j \neq i$ we find that the events $\{y_j \leftrightarrow \infty \text{ in } S_j\}$ are $\mathcal{F}_{E_\infty}$ -measurable for $j \neq i$. The estimate (3.17) holds for each choice of i. Therefore

$$\mu^*_{p,d}\{y_i \leftrightarrow \infty \text{ in } S_i \text{ for at least one } i \leq \nu|\mathcal{F}_s\} \geq 1 - \{1 - C_2(\bar{\gamma})\}^\nu .$$

(3.11), (3.12)-(3.14) and (3.18) together yield (for large enough $d$ and $n$)

$$\begin{aligned}\mu^*_{p,d}\{B(x,n)\} \leq & (1 - c_1(\gamma'',\ell))^n \\ & + \mu^*_{p,d}\left\{s\nu_s(x) > (2\gamma')^{-1}(\gamma'-1)\,d \text{ for some } s \leq 2n\right\} \\ & + (1 - C_2(\bar{\gamma}))^\ell .\end{aligned} \tag{3.19}$$

The second term on the right hand side is at most

$$\begin{aligned} & \mu^*_{p,d}\left\{|\Delta_{2n}(x)| > \tfrac{1}{n}(\gamma')^{-1}(\gamma'-1)\,d\right\} \\ & \leq \mu^*_{p,d}\left\{\Delta_{2n}(x) \neq \tilde{\Delta}_{2n}(x)\right\} \\ & \quad + O\left(\tfrac{1}{d}E^*_{p,d}\left\{\left|\tilde{\Delta}_{2n}(x)\right|\right\}\right) \\ & = O\left(\tfrac{1}{d}\right)\end{aligned} \tag{3.20}$$

for fixed $n$ and $p \leq \gamma''(2d)^{-1}$, by virtue of (3.7), (3.8). (3.9) now follows by first choosing $\ell$ large, and then letting $d \to \infty$ and $n \to \infty$ (in that order).

*Remark 4.* The restriction $p \leq \gamma''(2d)^{-1}$ in (3.9) can be removed, but we shall not need this.

Define

$$J(x) = \text{ indicator function of } \{x \leftrightarrow \infty\}.$$

**Lemma 8.** *Let $\gamma'' \geq \gamma' > 1$ and $k$ be fixed. Then*

$$\begin{aligned}\mu^*_{p(d),d}\{\, I(x) \neq J(x)\} \to 0 \quad (d \to \infty) \quad & \textit{uniformly in} \\ & \left\{|x| \leq k, Q\gamma'(2d)^{-1} \leq p(d) \leq \gamma''(2d)^{-1}\right\}\end{aligned} \tag{3.21}$$

*(see (3.1) for $I(\cdot)$),*

$$\text{(3.22)}\qquad \begin{aligned}&\mu^*_{p(d),d}\{x' \leftrightarrow x'' \text{ but } x' \not\leftrightarrow \infty\} \to 0 \quad (d\to\infty) \text{ uniformly in}\\ &\qquad\left\{|x'|,|x''| \le k, Q\gamma'(2d)^{-1} \le p(d) \le \gamma''(2d)^{-1}\right\},\end{aligned}$$

*and for any fixed $\gamma''$ there exists a $C_3(\gamma'') < \infty$ such that*

$$\text{(3.23)}\qquad \begin{aligned}0 \le \Big[&E^*_{p(d),d}\{J(x')J(x'')\} - E^*_{p(d),d}\{J(x')\}\,E^*_{p(d),d}\{J(x'')\}\Big]\\ &\le \mu^*_{p(d),d}\{B(x'',n)\} + C_3(\gamma'')(2n+1)(\gamma'')^{2n}d^{-1}\\ &\quad + \Big[\mu^*_{p(d),d}\{0\leftrightarrow\infty\} - \mu^*_{p(d),d-1}\{0\leftrightarrow\infty\}\Big],\end{aligned}$$

*for all $n$ and $p(d) \le \gamma''(2d)^{-1}$.*

*Proof.* For brevity we write $\mu^*_d$ instead of $\mu^*_{p(d),d}$, and $E^*_d$ for expectation with respect to $\mu^*_{p(d),d}$ in this proof.

We start with (3.21). Clearly $J(x) \ge I(x)$, and $J(x) > I(x)$ only if $x \leftrightarrow \infty$, but only via paths through the origin. Write

$$J_n(x) = \text{ indicator function of } \{x \leftrightarrow \partial\Lambda_n\}.$$

Then

$$\text{(3.24)}\qquad \begin{aligned}\mu^*_d\{J(x) > I(x)\} \le \mu^*_d\{&x\leftrightarrow \mathbf{0} \text{ in } \Lambda_{2n}\}\\ &+\{x\leftrightarrow\infty \text{ only via paths which go from } x\\ &\qquad \text{to } \Lambda^c_{2n} \text{ without passing through } \mathbf{0},\\ &\qquad \text{and then back to } \mathbf{0} \text{ and from there to } \infty\}.\end{aligned}$$

The event in the second term in the right hand side of (3.24) can occur only if $x \leftrightarrow \partial\Lambda_{2n}$ but there is no $s \in (n, 2n]$ and $y \in \Lambda_s$ such that $x \leftrightarrow y$ in $\Lambda_s\backslash\{\mathbf{0}\}$ and $y \leftrightarrow \infty$ outside $\Lambda_s$. Thus the second term in the right hand side of (3.24) is bounded by

$$\text{(3.25)}\qquad \mu^*_d\{x\leftrightarrow \mathbf{0} \text{ in } \Lambda_{2n}\} + \mu^*_d\{B(x,n)\}$$

(see Lemma 9 for $B(x,n)$). Notice also that if $x = \pm\xi_{i_1} \pm \xi_{i_2} \pm \ldots \pm \xi_{i_r}, r \leq k$, then

$$\begin{aligned}
\mu_d^*\{x \leftrightarrow 0 \text{ in } \Lambda_{2n}\} &= \mu_d^*\{0 \in \Delta_{2n}(x)\} \\
&\leq \mu_d^*\{\Delta_{2n}(x) \text{ contains a vertex } z \text{ such that} \\
&\qquad\qquad \text{the edge from } z \text{ to } z \pm \xi_{i_1} \text{ is occupied}\} \\
&\leq p(d)\,(2n+1) \sum_{\lambda=0}^{2n} (2dp(d))^{\lambda}
\end{aligned} \tag{3.26}$$

(see Lemma 6). Finally, for any $\varepsilon > 0$ we can (by (3.9)) fix $n$ such that

$$\limsup_{d\to\infty} \sup \left\{\mu_{pd}^*\{B(x,n)\} : |x| \leq k, Q\gamma'\,(2d)^{-1} \leq p \leq \gamma''\,(2d)^{-1}\right\} \leq \varepsilon.$$

From the above we then obtain

$$\begin{aligned}
\limsup_{d\to\infty} \mu_{p(d),d}^*\{I(x) \neq J(x)\} &\leq 2 \limsup_{d\to\infty} p(d)\,(2n+1) \sum_{\lambda=0}^{2n} (2dp(d))^{\lambda} + \varepsilon \\
&= \varepsilon,
\end{aligned}$$

and this estimate is uniform in $\left\{|x| \leq k, q\gamma'\,(2d)^{-1} \leq p(d) \leq \gamma''\,(2d)^{-1}\right\}$. This proves (3.21).

A very similar argument shows that the left hand side of (3.22) is bounded by

$$\mu_{p(d),d}^*\{x' \leftrightarrow x'' \text{ in } \Lambda_{2n}\} + \mu_{p(d),d}^*\{B(x',n)\},$$

which can then be estimated in the same way as (3.25). Thus (3.22) also holds.

Next we turn to (3.23). First we note that the covariance of $J(x')$ and $J(x'')$ is positive by the FKG inequality (see [ACCN, Sect. 2.2]); formally we can apply this inequality only to increasing cylinder functions, so we apply it to $J_n(x')$ and $J_n(x'')$ and then let $n$ go to $\infty$. To prove the right hand inequality in (3.23) we write the covariance of $J(x')$ and $J(x'')$ as

$$-E_d^*\{J(x')\,(1-J(x''))\} + E_d^*\{J(x')\}\,E_d^*\{1-J(x'')\}. \tag{3.27}$$

Now $1 - J(x'') \neq 0$ is equivalent to the cluster of $x'', \mathcal{C}(x'')$, being finite. Therefore

$$\{1 - J(x'') \neq 0\} = \bigcup_C \{\mathcal{C}(x'') = C\}$$

where the union in the right hand side is over all finite connected sets of edges, which contain $x''$. This union over $C$ is clearly adisjoint union, since $\{\mathcal{C}(x'') = C_1\} \cap \{\mathcal{C}(x'') = C_2\} = \emptyset$ whenever $C_1 \neq C_2$. The expression in (3.27) can therefore be rewritten as

$$\text{(3.28)} \quad \sum_C \mu_d^* \{\mathcal{C}(x'') = C\} \left[-\mu_d^* \{x' \leftrightarrow \infty | \mathcal{C}(x'') = C\} + \mu_d^* \{x' \leftrightarrow \infty\}\right].$$

We break the sum over $C$ up into three pieces. The first sum, $\sum_1$ say, contains those $C$ which intersect $\Lambda_{2n}^c$ = complement of $\Lambda_{2n}$. This sum is at most

$$\mu_d^* \{x'' \leftrightarrow \partial\Lambda_{2n} \text{ but } x'' \not\leftrightarrow \infty\} \leq \mu_d^* \{B(x'', n)\}.$$

To describe the next two subsums we write $x' - x''$ as $a_1\xi_{i_1} + \quad a_2\xi_{i_2} + \ldots + a_r\xi_{i_r}$ for some (unique) $r, i_s$, and nonzero integers $a_i$. For the sake of argument let $a_1 \geq 1$. We then take for $\sum_2$ the sum over all $C$ which are contained in $\Lambda_{2n}$ and which contain an edge from some $z$ to $z + \xi_{i_1}$. The subsum of (3.28) over these $C$ is as most

$$\begin{aligned} &\mu_d^* \{\Delta_{2n}(x'') \text{ contains a vertex } z \text{ such that} \\ &\qquad\qquad \text{the edge from } z \text{ to } z + \xi_{i_1} \text{ is occupied}\} \\ &\leq p(d)\,(2n+1) \sum_{\lambda=0}^{2n} (p(d) \cdot 2d)^\lambda \\ &\leq C_3\,(2n+1)\,(\gamma'')^{2n}\, d^{-1} \end{aligned}$$

for some constant $C_3 = C_3(\gamma'')$, whenever $p(d) \leq \gamma''(2d)^{-1}$ (see (3.8)). Finally, $\sum_3$ contains all $C \subset \Lambda_{2n}$ which are not contained in $\sum_2$.

To estimate the sum $\sum_3$ we consider the subspace $S := span\{\xi_j : j \neq \xi_{i_1}\}$, and its translate $x' + S$. We define $\tilde{C}$ as the "boundary of $C$", i.e.,

$$\tilde{C} = \{\text{all edges incident to } C, \text{ but not in } C\}.$$

Then, for any $C$ contained in $\sum_3, \tilde{C}$ contains no edge in $x' + S$. Indeed, if $z$ is a vertex in $C$, then its $i_1$-th coordinate, $z(i_1)$, cannot be more than $x''(i_1)$, the $i_1$-th coordinate of $x''$, because there are no paths in $C$ with a step in the positive $i_1$-direction. On the other hand, any vertex in $x'+S$ has $i_1$ coordinate equal to $x'(i_1) = x''(i_1) + a_1 > x''(i_1)$. We also note that

$$\begin{aligned}&\{\mathcal{C}\,(x'') = C\}\\ &\quad = \{\text{all edges in } C \text{ are occupied}\} \cap \{\text{all edges in } \tilde{C} \text{ are vacant}\}.\end{aligned}$$

Given that all edges in $\tilde{C}$ are vacant, the edges in $C$ can have no influence on the components of points outside $C$. Therefore, for any bounded cylinder function $f$ which depends only on edges outside $C$ we have

$$E_d^*\{f|\mathcal{C}\,(x'') = C\} = E_d^*\{f| \text{ all edges in } \tilde{C} \text{ are vacant}\}.$$

We now apply this observation, together with Lemma 1 and Cor. 2, to the indicator function of $\{x' \leftrightarrow \partial\Lambda_m \text{ in } x' + S\}$ for fixed $m$. We then obtain, very much as in the proof of (3.16),

$$\begin{aligned}&\mu_d^*\{x' \leftrightarrow \infty \mid \mathcal{C}\,(x'') = C\}\\ &\quad \geq \mu_d^*\left\{x' \leftrightarrow \partial\Lambda_m \text{ in } x' + S \mid \text{all edges in } \tilde{C} \text{ are vacant}\right\}\\ &\quad \geq \mu_d^*\{x' \leftrightarrow \partial\Lambda_m \text{ in } x' + S \mid \text{all edges not in } x' + S \text{ are vacant}\}\\ &\quad \geq \mu_{p(d),d-1}^*\left\{0 \leftrightarrow \partial\Lambda_{m-|x'|}\right\}.\end{aligned}$$

In the last step we used that $S$ is isomorphic to $\mathbb{Z}^{d-1}$. Finally, upon letting $m \to \infty$ we obtain for any $C$ in $\sum_3$

$$\mu_d^*\{x' \leftrightarrow \infty|\mathcal{C}\,(x'') = C\} \geq \mu_{p(d),d-1}^*\{0 \leftrightarrow \infty\}$$

(3.23) follows by combining the estimates for $\sum_1 - \sum_3$.

*Proof of Theorem 2.* We are finally ready to show that the covariance of $J(x')$ and $J(x'')$ tends to zero as $d \to \infty$ for suitable $p(d)$. Specifically, for the remainder of this paper $p(d) = Q\gamma\,(2d)^{-1}$ with $\gamma > 1$ independent of $d$. We also fix $*$ as $f$ or $w$ and $Q \in [1,2]$. Define

$$\bar{\theta}(\gamma) = \limsup_{d\to\infty} \theta^*\left(\frac{Q\gamma}{2d}, \mathbb{Z}^d, Q\right)$$
$$\underline{\theta}(\gamma) = \liminf_{d\to\infty} \theta^*\left(\frac{Q\gamma}{2d}, \mathbb{Z}^d, Q\right).$$

$\bar{\theta}$ and $\underline{\theta}$ are increasing, bounded functions of $\gamma$, and therefore have at most countably many points of discontinuity. Now let $\gamma$ be a continuity point of $\bar{\theta}$ and of $\underline{\theta}$ and $p(d) = Q\gamma\,(2d)^{-1}$. Then, by (3.23) and (3.9), for any $|x'| = |x''| = 1, x' \neq x''$, and for any subsequence $\mathcal{D} = \{d_1 < d_2 < \dots\}$

$$\begin{aligned}
&\limsup_{\substack{d\to\infty \\ \text{in } \mathcal{D}}} \left[E^*_{p(d),d}\{J(x')J(x'')\} - E^*_{p(d),d}\{J(x')\}\,E^*_{p(d),d}\{J(x'')\}\right] \\
&\quad\leq \lim_{n\to\infty} \limsup_{d\to\infty} \mu^*_{p(d),d}\{B(x'',n)\} \\
&\qquad + \limsup_{\substack{d\to\infty \\ \text{in } \mathcal{D}}} \left[\mu^*_{p(d),d}\{0 \leftrightarrow \infty\} - \mu^*_{p(d),d-1}\{0 \leftrightarrow \infty\}\right] \qquad (3.29)\\
&\quad= \limsup_{\substack{d\to\infty \\ \text{in } \mathcal{D}}} \left[\mu^*_{p(d),d}\{0 \leftrightarrow \infty\} - \mu^*_{p(d),d-1}\{0 \leftrightarrow \infty\}\right].
\end{aligned}$$

Now first take a subsequence $\mathcal{D}_1$ such that

$$\limsup_{\substack{d\to\infty \\ \text{in } \mathcal{D}_1}} \mu^*_{p(d),d}\{0 \leftrightarrow \infty\} = \limsup_{\substack{d\to\infty \\ \text{in } \mathcal{D}_1}} \theta^*\left(\frac{Q\gamma}{2d}, \mathbb{Z}^d, Q\right) = \underline{\theta}(\gamma).$$

Then for any $\epsilon > 0$

$$p(d) = \frac{Q\gamma}{2d} \geq \frac{Q\,(\gamma - \varepsilon)}{2\,(d-1)} \quad \text{for large } d,$$

and therefore

$$\liminf_{\substack{d\to\infty \\ \text{in } \mathcal{D}_1}} \mu^*_{p(d),d-1}\{0 \leftrightarrow \infty\} \geq \liminf_{d\to\infty} \theta^*\left(\frac{Q\,(\gamma-\varepsilon)}{2\,(d-1)}, \mathbb{Z}^{d-1}, Q\right) = \underline{\theta}(\gamma - \epsilon).$$

Thus, for $\mathcal{D} = \mathcal{D}_1$ the last member of (3.29) is at most $\underline{\theta}(\gamma) - \underline{\theta}(\gamma - \epsilon)$ for any $\epsilon > 0$. Since $\gamma$ was taken as a continuity point of $\underline{\theta}$ this implies

$$\text{(3.30)} \quad \limsup_{\substack{d\to\infty \\ \text{in } \mathcal{D}_1}} \left[E^*_{p(d),d}\{J(x')J(x'')\} - E^*_{p(d),d}\{J(x')\}\, E^*_{p(d),d}\{J(x'')\}\right] = 0$$

for any $|x'| = |x''| = 1, x' \neq x''$. In fact the covariance in square brackets has the same value for all such $x', x''$ so that (3.30) is uniform in such pairs $x', x''$. It follows from (3.21) that

$$\text{(3.31)} \qquad \frac{1}{2d}M - \frac{1}{2d}\sum_{|x|=1} J(x) \to 0 \text{ in distribution as } d \to \infty.$$

Moreover, as $d \to \infty$ in $\mathcal{D}_1$, (3.30) implies

$$\text{(3.32)} \qquad \text{variance}^*_{p(d),d}\left\{\frac{1}{2d}\sum_{|x|} J(x)\right\} \to 0,$$

and by our choice of $\mathcal{D}_1$

$$\text{(3.33)} \qquad E^*_{p(d),d}\left\{\frac{1}{2d}J(x)\right\} = \theta^*\left(\frac{Q\gamma}{2d}, \mathbb{Z}^d Q\right) \to \underline{\theta}(\gamma).$$

We now take $p = Q\gamma(2d)^{-1}$ in (3.4) and $\beta = -log(1-p) \sim Q\gamma(2d)^{-1}$. If we now let $d \to \infty$ in $\mathcal{D}_1$, then the left hand side of (3.4) tends to $\underline{\theta}(\gamma)$, while, by virtue of (3.31) - (3.33) the right hand side converges to $f_Q(Q\gamma\underline{\theta}(\gamma))$, where

$$f_Q(x) = \frac{1 - \exp(-x)}{1 + (Q-1)\exp(-x)}$$

We have therefore proved that for any continuity point $\gamma$ of $\underline{\theta}(\cdot)$, $\underline{\theta}(\gamma)$ is a solution of

$$\text{(3.34)} \qquad \underline{\theta}(\gamma) = f_Q(Q\gamma\underline{\theta}(\gamma)).$$

Moreover, we already know from (3.10) that $\underline{\theta}(\gamma)$ must be strictly positive for $\gamma > 1$. Finally $x \to f_Q(x)$ is concave for $1 \leq Q \leq 2$, so that (3.34) has a unique strictly positive solution. Thus $\underline{\theta}(\gamma)$ is the unique strictly positive solution of (3.34) for all but countably many $\gamma > 1$. In fact this must be true for *all* $\gamma > 1$ since $\underline{\theta}(\gamma)$ is increasing in $\gamma$, and the solution of $\theta = f_Q(Q\gamma\theta)$ is continuous in $\gamma$ (by the implicit function theorem).

We can now appeal to the concavity of $f_Q(\cdot)$ to obtain from (3.4) that along any sequence $\mathcal{D}$ for which

$$\hat{\theta} := \lim_{\substack{d\to\infty \\ \text{in } \mathcal{D}}} \theta^*\left(\frac{\gamma Q}{2d}, \mathbb{Z}^d, Q\right) \text{ exists,} \tag{3.35}$$

one has, by Jensen's inequality,

$$\hat{\theta} \le f_Q\left(\gamma Q \lim E^*_{p(d),d}\left\{\tfrac{1}{2d}M\right\}\right) = f_Q\left(\gamma Q\hat{\theta}\right). \tag{3.36}$$

In particular $\bar{\theta} \le f_Q(\gamma Q\bar{\theta}(\gamma))$. Since $f_Q(\gamma Qx) < x$ for $x > \underline{\theta}(\gamma)$ this shows $\bar{\theta}(\gamma) \le \underline{\theta}(\gamma)$ and hence $\bar{\theta}(\gamma) = \underline{\theta}(\gamma)$. Thus, for $\gamma > 1$

$$\lim_{d\to\infty} \theta^*\left(\frac{\gamma Q}{2d}, \mathbb{Z}^d, Q\right)$$

exists and is the unique strictly positive solution of (3.34). This proves (1.5) and (1.6).

The last argument does not work for $Q > 2$, because $f_Q$ is no longer concave in that case. Some comments about this are given in Remark 5 below. We first complete the proof for $1 \le Q \le 2$, though.

Once we have the existence of the limit in (1.5) and (1.6) we can go back to (3.29). We now see that the last member of (3.29) vanishes even for $\mathcal{D}$ equal to the full sequence of all positive integers, and hence the same holds for its first member. Thus also (3.32) holds without restriction on $d$, so that

$$\frac{1}{2d}M - y(\gamma) \to 0 \text{ is distribution } (d \to \infty)$$

(for $p(d) = \gamma Q(2d)^{-1}, \gamma > 1$). This replaces equation (2.21) of [BKLS], and the proof of (1.7) is now virtually the same as that of (1.10) in [BKLS].

Finally, (1.8) is contained in (2.11), (2.16) and (2.17).

*Remark 5.* The arguments based on the concavity of $f_Q(\cdot)$ do not carry over to $Q > 2$. The following alternative argument to replace (3.35), (3.36) is therefore of some interest. Observe that for any $p, \ell, m > 0$ (compare (3.16))

$$\begin{aligned}\mu^*_{p,d-\ell}\{0 \leftrightarrow \partial\Lambda_m\} = \lim_{n\to\infty} \mu^*_{p,d}\{0 \leftrightarrow \partial\Lambda_m \text{ in span } \{\xi_1, \dots, \xi_{d-\ell}\} \mid \\ \omega(e) = 0 \text{ for all } e \text{ in } \Lambda_n \\ \text{outside span } \{\xi_1, \dots, \xi_{d-\ell}\}\} \\ \leq \mu^*_{p,d}\{0 \leftrightarrow \partial\Lambda_m\}\end{aligned}$$

Thus, by letting $m \to \infty$ we see that $\mu^*_{p,d}\{0 \leftrightarrow \infty\}$ is increasing in $d$. Now let $\mathcal{D}_2$ be any subsequence such that

$$\lim_{\substack{d\to\infty \\ \text{in } \mathcal{D}_2}} \theta^*\left(\frac{\gamma Q}{2(d-1)}, \mathbb{Z}^{d-1}, Q\right) \geq \bar{\theta}(\gamma).$$

Then, if $\gamma$ is a continuity point of $\bar{\theta}(\gamma)$

$$\lim_{\substack{d\to\infty \\ \text{in } \mathcal{D}_2}} \theta^*\left(\frac{\gamma Q}{2(d-1)}, \mathbb{Z}^{d}, Q\right) \geq \bar{\theta}(\gamma),$$

and by definition of $\bar{\theta}$, the last inequality must be an equality. The argument for (3.30) - (3.34) can now be repeated almost verbatim, with $\mathcal{D}_1$ and $\underline{\theta}$ replaced by $\mathcal{D}_2$ and $\bar{\theta}$, respectively. It follows that $\bar{\theta}$ also must be a solution of (3.34). As before, for $Q \leq 2$ this shows $\underline{\theta} = \bar{\theta}$.

For $Q > 2$ (3.34) actually has, for a certain interval of $\gamma$'s, two strictly positive solutions, $0 < \rho_-(\gamma) < \rho_+(\gamma)$ say (cf. [KS], Sect. 3). However, $\rho_-(\gamma)$ is strictly decreasing function of $\gamma$, while $\underline{\theta}(\cdot)$ and $\bar{\theta}(\cdot)$ are increasing. From this one concludes that there can be at most one value $\gamma_0$ for which $\underline{\theta}(\gamma_0) = \rho_-(\gamma_0) > 0$. For all $\gamma > \gamma_0$, $\underline{\theta}(\gamma)$ and $\bar{\theta}(\gamma)$ must then equal $\rho_+(\gamma)$, the largest positive solution of $x = f_Q(Q\gamma x)$.

## References

[ACCN] Aizenman, M, Chayes, J.T., Chayes, L. and Newman, C.M., *Discontinuity of the magnetization in one-dimensional* $l/|x-y|^2$ *Ising and Potts models*, J. Stat. Phys. **50** (1988), pp. 1–40.

[AKN] Aizenman, M., Kesten, H. and Newman, C.M., *Uniqueness of the infinite cluster and continuity of connectivity functions for short and long range percolation*, Comm. Math. Phys. **111** (1987), pp. 505-531.

[BB] van den Berg, J. and Burton, R., *FKG and equivalent conditions for binary random variables*, preprint (1987).

[BK] Burton, R.M. and Keane, M., *Density and uniqueness in percolation*, Comm. Math. Phys. **121** (1989), pp. 501–505.

[BKLS] Bricmont, J., Kesten, H., Lebowitz, J. and Schonmann, R.H., *A note on the large dimensional Ising model*, Comm. Math. Phys. **122** (1989), pp. 597–607.

[ES] Edwards, R. and Sokal, A., *Generalization of the Fortuin-Kasteleyn-Swendsen-Wang representation and Monte Carlo algorithms*, Phys. Rev. D **38** (1988), pp. 2009–2012.

[Fi] Fisher, M.E., *Critical temperatures of anisotropic Ising lattices. III. General upper bounds*, Phys. Rev. **162** (1967), pp. 480–485.

[Fo] Fortuin, C.M., On the random cluster model III. The simple random cluster model, Physica **59** (1972), pp. 545–570.

[GNK] Gandolfi, A. Keane, M.S. and Newman, C.M., *Uniqueness of the infinite component in a random graph with applications to percolation and spin glasses*, preprint, Delft University of Technology (1989).

[GGR] Gandolfi, A., Grimmett, G. and Russo, L., *On the uniqueness of the infinite cluster in the percolation model*, Comm. Math. Phys. **114** (1988), pp. 549–552.

[Ha] Harris, T.E., *A lower bound for the critical probability in a certain percolation process*, Proc. Cambr. Phil. Soc. **56** (1960), pp. 13–20.

[Ho] Holley, R., *Remarks on the FKG inequalities*, Comm. Math. Phys. **36**, pp. 227–231.

[HS] Hara, T. and Slade, G., *Mean-field critical phenomena for percolation in high dimensions*, Comm. Math. Phys. **128** (1990), pp. 333–391.

[Km] Kemperman, J.H.B., *On the FKG-inequality for measures on a partially ordered space*, Proc. Kon. Ned. Akad. van Wetensch., Ser. A **80** (1977), pp. 313–331.

[Ks] Kesten, H., *Asymptotics in high dimension for percolation*, in "Disorder in Physical Systems, A Volume in Honor of J.M. Hammersley," eds. G. R. Grimmett and D. J. A. Welsh, Oxford University Press, 1990, pp. 219–240.

[KS] Kesten, H. and Schonmann, R., *Behavior in large dimensions of the Potts and Heisenberg models*, Rev. Math. Phys. **1** (1990), pp. 147–182.

[SV] Schonmann, RH. and Vares, M.E., *The survival of the large dimensional basic contact process*, Prob. Th. Rel. Fields **72** (1986), pp. 387–393.

Research supported by the NSF through a grant to Cornell University.

Department of Mathematics
Cornell University
Ithaca, NY 14853

Interacting Particle Systems

# On the Asymptotics of the Spin-Spin Autocorrelation Function In Stochastic Ising Models Near the Critical Temperature

Richard Holley

*Abstract.* The rate at which the spin-spin autocorrelation function goes to zero as the time goes to infinity in stochastic ISING models with ferro-magnetic interactions and attractive flip rates is investigated. It is first shown that if this rate is exponential then the exponent is the same as the gap between 0 and the rest of the spectrum of the infinitesimal generator as an operator on the $L^2$ space of the equilibrium measure. It is easy to get upper bounds on the gap in the spectrum. The best known such bounds are related to the susceptibility of the magnetization. In two dimensions the spin-spin autocorrelation function is investigated when the temperature is exactly at the critical temperature. The rate of convergence to 0 is no longer exponential in this case, and again we obtain only upper bounds. No lower bounds are known at any temperatures near the critical temperature.

## §0 Introduction

This paper resulted from an attempt by the author to understand the nature of the critical slowing down of stochastic ISING models at their critical temperatures. There are several papers in the physics literature on this subject (see [B], [K] [A&H1] [B&K] [H&H] and the references therein). Many of these papers are reports on numerical experiments involving simulations of some particular version of a stochastic ISING model on moderate sized lattices. Some of the above papers contain results of the form: the rate of convergence, when measured in a certain way, is no faster than .... The upper bounds on the rates are usually given in terms of various critical exponents of the GIBBS states which are the stationary measures of the processes. The approach that we take here is to measure the rate of convergence in terms of the size of the gap between 0 and the rest of the spectrum of the infinitesimal generator of the process. This is closely related to, but as far as the author can tell, not exactly the same as any of the methods used in the above cited papers. The spectral gap gives us a lower bound on the rate at which the semi-group acting on any $L^2$ function converges to its equilibrium value. The main purpose of this paper is to identify the $L^2$ functions on which the convergence is as slow as possible. Those functions turn out to be the functions that simply evaluate the configuration of spins at a single site. That these functions are the one on which the convergence is the slowest in the sense of $L^2$ convergence is not at all surprising in view of the fact that for attractive spin flip systems the rate of convergence of the average magnetization controls the rate of convergence on all cylinder functions when the rate of convergence is measured in the uniform norm.

While it is easy to obtain upper bounds on the spectral gap, no lower bounds are known at temperatures near the critical temperature. The results in this paper may possibly be a small step in the direction of finding lower bounds on the spectral gap.

In this paper we deal exclusively with interactions which are translation invariant, finite range, ferro-magnetic pair potentials. Thus for every set $R \subset \mathbf{Z}$ with one or two elements we have a number $J_R$ and we assume that the $J_R$'s have the property that $J_R = J_{R+k}$ for all $k \in \mathbf{Z}$ and that $J_R = 0$ if $\mathrm{Diam}(R) > L$ ($L$ is the range of the interaction).

Next, set $\mathbf{E} = \{-1,1\}^{\mathbf{Z}}$ and think of the elements $\sigma$ of $\mathbf{E}$ as configurations on $\mathbf{Z}$ of $\pm 1$ valued spins. Thus, $\sigma_k \in \{-1,1\}$ is the spin at site $k \in \mathbf{Z}$ of the configuration $\sigma \in \mathbf{E}$. Also, given a finite, non-empty set $R \subset \mathbf{Z}$, we will use the notation

$$\sigma_R \equiv \prod_{i \in R} \sigma_i .$$

The GIBBS states determined by the interaction $\{J_R : R \subset \mathbf{Z}\}$ are the probability measures, $\mu$, on $\mathbf{E}$ whose finite dimensional conditional distributions are given by

$$\mu\big(\sigma_i = \omega_i,\, i \in \Lambda \,|\sigma_j = \eta_j,\, j \in \Lambda^c\big) = \frac{\exp\left[\sum_{R \subset \Lambda} J_R \omega_R + \sum_{R\cap\Lambda^c \neq \emptyset, R\cap\Lambda \neq \emptyset} J_R \omega_{R\cap\Lambda} \eta_{R\cap\Lambda^c}\right]}{Z(\Lambda,\eta)}$$

for any finite set $\Lambda$. Here $Z(\Lambda,\eta)$ is the normalizing constant needed to make the expression on the right-hand side a probability measure. We assume throughout this paper that the interaction $\{J_R : R \subset \mathbf{Z}\}$ is such that there is only one GIBBS state, which we denote by $\mu$, for that interaction.

The *flip rates* of a stochastic ISING model are a family $\{c_k : k \in \mathbf{Z}\}$ of functions $c_k : \mathbf{E} \longrightarrow (0,\infty)$ which satisfy the *detailed balance condition*:

$$c_k(\sigma)\exp\left[\sum_{R\ni k} J_R\sigma_R\right] = c_k(\sigma^k)\exp\left[-\sum_{R\ni k} J_R\sigma_R\right],$$

where $\sigma^k$ is the configuration of spins that agrees with $\sigma$ except at $k$, at which site the spin is $-\sigma_k$. We will be assuming throughout that, in addition, the $c_k$*'s* are *translation invariant*, have *finite range* $L \in \mathbf{Z}^+$, and are attractive. That is, $c_k(\sigma) = c_j(\tau)$ if $\sigma_i = \tau_{j-k+i}$ for all $i \in Z$, $c_k(\sigma) = c_k(\omega)$ if $\sigma_\ell = \omega_\ell$ for all $\mid k-\ell \mid \leq L$, and $c_0(\sigma)$ is an increasing function of $\sigma$ on the set where $\sigma(0) = -1$ and a decreasing function of $\sigma$ on the set where $\sigma(0) = 1$. If the interaction is a ferro-magnetic pair potential (i.e. $J_R = 0$ if $|R| > 2$, $J_R \geq 0$ if $|R| = 2$ and $J_{\{0\}}$ is arbitrary) there are many ways to choose the flip rates to satisfy these conditions. For example one could take

$$c_0(\sigma) = \exp\left[-\sum_k J_{0,k}\sigma_0\sigma_k - J_{\{0\}}\sigma_0\right].$$

Given flip rates $\{c_k : k \in \mathbf{Z}\}$, we define the operator $\mathcal{L}$ on the space $\mathcal{F}$ of cylinder functions by

$$\mathcal{L}f(\sigma) = \sum_{k\in Z} c_k(\sigma)(f(\sigma^k) - f(\sigma)),\ \sigma \in E,\ \text{for } f \in \mathcal{F}.$$

The corresponding stochastic ISING model is the (unique) MARKOV process whose infinitesimal generator $\hat{\mathcal{L}}$ extends $\mathcal{L}$; and we denote by $\{T_t : t \geq 0\}$ the associated semigroup. The following facts are well known:

(1) $\mathcal{F}$ is a core for $\hat{\mathcal{L}}$,
(2) $\{T_t : t \geq 0\}$ is FELLER continuous (i.e. it takes $C(E;\mathbf{R})$ into itself) and has the GIBBS states $\mu$ as its stationary measures,
(3) If there is only one GIBBS state then it is the unique stationary measure for $\{T_t : t \geq 0\}$,
(4) For every GIBBS state, $\mu$, $\mathcal{L}$ is essentially self-adjoint on the space $L^2(\mu)$.

We will be dealing with functions which converge to zero as time goes to infinity and we want a way to measure the rate of convergence. We have chosen the following method. If $g(t) \to 0$ as $t \to \infty$ we say that $g$ converges to zero exponentially fast with exponent $\epsilon$ if

$$\lim_{t\to\infty} -\frac{1}{t}\log(g(t)) = \epsilon. \tag{0.1}$$

From the spectral theorem it follows (see Theorem (2.1) below) that if $g \in L^2(\mu)$ and $\int g d\mu = 0$ and there is only one GIBBS state then $\|T_t g\|_{L^2(\mu)}$ converges to zero exponentially fast with exponent at least as large as the gap between 0 and the rest of the spectrum of $\mathcal{L}$. We are looking for the $L^2$ function whose exponent is as small as possible. The content of Theorem (2.3) is that the desired function is $\phi_k(\sigma) = \sigma(k)$ for any $k \in \mathbf{Z}$. When measuring the rate of convergence in terms of the uniform norm instead of the $L^2$ norm the function to study is $T_t\phi_0(\vec{1}) - T_t\phi_0(\overrightarrow{-1})$ (see [L]). It is known that this quantity controls the rate of convergence of $T_t f(\sigma) - \int f(\eta)\mu(d\eta)$ to zero; however it is not known that it has an exponent associated with it, i.e. that the limit in (0.1) exists. That is the content of section 1. We include it to emphasize the similarity between the uniform and $L^2$ notions of convergence rates for the stochastic ISING model. Finally in Section 3 we consider the case when the model is exactly at the critical temperature. In that case the exponent is zero. All that a zero exponent tells us is that the convergence is not exponentially fast, and in this section we try to characterize the rate of convergence a little more precisely.

## §1 Single spins and the rate of convergence

In this section we investigate how the rate at which expectations of single spins converge to their equilibrium value is related to the rate at which expectations of all other cylinder functions converge to their equilibrium values. In the case of uniform convergence it is well known that the rate at which expectations of single spins converge controls the rate at which expectations of all cylinder functions converge. In fact the following theorem is known (see [L]).

**Theorem (1.1).** *For all $f \in \mathcal{F}$ there is a constant $A_f < \infty$ such that for all $\eta \in \mathbf{E}$ and all $t \geq 0$*

$$| T_t f(\eta) - \int f d\mu | \leq A_f \left(T_t\phi_0(\vec{1}) - T_t\phi_0(\overrightarrow{-1})\right)$$

Thus if we want to characterize the rate of convergence in the uniform norm it suffices to characterize the rate at which $T_t\phi_0(\vec{1}) - T_t\phi_0(\overrightarrow{-1})$ goes to zero. If this rate is exponential then we may obtain the exponent from

$$\lim_{t\to\infty} \frac{-1}{t} \log(T_t\phi_0(\vec{1}) - T_t\phi_0(\overrightarrow{-1})). \tag{1.2}$$

Our next result says that the limit in (1.2) always exists; hence in view of Theorem (1.1) when the limit in (1.2) is positive it provides, to first order, a measurement of the rate of convergence to equilibrium. Note that the following theorem assumes that the flip rates have finite range and are attractive, but not that they satisfy detailed balance for some potential.

**Theorem (1.3).** *For any finite range attractive spin system*

$$\lim_{t\to\infty} \frac{-1}{t} \log(T_t\phi_0(\vec{1}) - T_t\phi_0(\overrightarrow{-1}))$$

*exists and is in* $[0, \infty)$.

*Proof.* By [A&H2, Theorem (2.12)] there are two possibilities.

1) For some $c > 0$ and all $t \geq 1$, $T_t\phi_0(\vec{1}) - T_t\phi_0(\overrightarrow{-1}) \geq ct^{-d}$, in which case

$$\frac{-1}{t} \log(T_t\phi_0(\vec{1}) - T_t\phi_0(\overrightarrow{-1})) \leq \frac{d\log(t) - \log(c)}{t}.$$

On the other hand, $T_t\phi_0(\vec{1}) - T_t\phi_0(\overrightarrow{-1}) \leq 2$ for all $t \geq 0$. Thus in this case

$$\lim_{t\to\infty} \frac{-1}{t} \log(T_t\phi_0(\vec{1}) - T_t\phi_0(\overrightarrow{-1})) = 0.$$

2) There exists $\gamma > 0$ and $t_0$ such that if $t \geq t_0$ then $T_t\phi_0(\vec{1}) - T_t\phi_0(\overrightarrow{-1}) \leq e^{-\gamma t}$. We let $\xi^t_{j,k} = \sup\{|T_t\phi_j(\sigma^k) - T_t\phi_j(\sigma)| : \sigma \in \mathbf{E}\}$, and note that

$$\xi^t_{j,k} \leq T_t\phi_0(\vec{1}) - T_t\phi_0(\overrightarrow{-1}), \tag{1.4}$$

$$\xi^{s+t}_{j,k} \leq \sum_{\ell} \xi^s_{j,\ell}\xi^t_{\ell,k}, \tag{1.5}$$

and

$$T_t\phi_0(\vec{1}) - T_t\phi_0(\overrightarrow{-1}) \leq \sum_k \xi^t_{0,k}. \tag{1.6}$$

For a proof of these inequalities see [A&H2]. Define $\delta(t) = \sum_k \xi^t_{0,k}$. From (1.5) we have $\delta(t+s) \leq \delta(t)\delta(s)$. Therefore $-\log(\delta(t))$ is super-additive, and hence

$$\lim_{t\to\infty} \frac{-1}{t}\log(\delta(t)) = \sup_{t>0} \frac{-1}{t}\log(\delta(t)) \leq \infty$$

exists. By the standard coupling (see [L]) it is easily seen that $\xi^t_{0,0} \geq e^{-2Ct}$, where $C = \sup_{\eta\in\mathbf{E}} c_0(\eta)$. Therefore $\lim_{t\to\infty} \frac{-1}{t}\log(\delta(t)) \leq 2C$. Let $\lim_{t\to\infty} \frac{-1}{t}\log(\delta(t)) = \delta$. Note that at this point we do not yet know that $\delta > 0$; but in any case from (1.6) we have

$$\varliminf_{t\to\infty} \frac{-1}{t}\log(T_t\phi_0(\vec{1}) - T_t\phi_0(\overrightarrow{-1})) \geq \delta.$$

Now by Theorem (2.11) in [H&S]

$$\sum_{|j|\geq 2MN} \xi^t_{j,0} \leq 6(1+M+2MN)^d \frac{t^{N+1}}{(N+1)!},$$

where $M$ is the range of the interaction. Therefore from (1.4) we have

$$\delta(t) \leq (T_t\phi_0(\vec{1}) - T_t\phi_0(\overrightarrow{-1}))(2MN+1)^d + 6(1+M+2MN)^d \frac{t^{N+1}}{(N+1)!}.$$

Let $N = [t\log(t)]$ (the greatest integer in $t\log(t)$). Then since $T_t\phi_0(\vec{1}) - T_t\phi_0(\overrightarrow{-1}) \geq \xi^t_{0,0} \geq e^{-2Ct}$, and since by Stirlings formula

$$\frac{t^{t\log(t)+1}}{([t\log(t)]+1)!} \leq \left(\frac{e}{\log(t)}\right)^{t\log(t)+1} = o(T_t\phi_0(\vec{1}) - T_t\phi_0(\overrightarrow{-1})),$$

it follows that

$$\delta(t) \leq 6(1+M+2Mt\log(t))^d(T_t\phi_0(\vec{1}) - T_t\phi_0(\overrightarrow{-1}))(1+o(1)).$$

Thus

$$0 < \gamma \leq \varlimsup_{t\to\infty} \frac{-1}{t}\log(T_t\phi_0(\vec{1}) - T_t\phi_0(\overrightarrow{-1})) \leq \delta.$$

Q.E.D.

## §2 Autocorrelation Functions

Throughout the rest of this paper we assume that we are dealing with an interaction for which the GIBBS state is unique. In that case it is known that 0 is an eigenvalue for the generator of the stochastic ISING model and the corresponding eigenvectors are the constants.

In this section we consider a stochastic ISING model $\{\sigma(t) : t \geq 0\}$ which is in equilibrium. I.e. if $\mu$ is the (unique) GIBBS state then the distribution of $\sigma(t)$ is $\mu$ for all $t \geq 0$. We then define the autocorrelation function of a single spin to be $R(t) = \int T_t\phi_0(\sigma)\phi_0(\sigma)\mu(d\sigma) - (\int \phi_0(\sigma)\mu(d\sigma))^2$

**Theorem (2.1).** *For all stochastic* ISING *models i.e. all choices of flip rates satisfying our assumptions,*

$$\lim_{t\to\infty} \frac{-1}{t} \log(R(t)) \tag{2.2}$$

*exists and is finite.*

*Proof.*
Let $\bar{\phi}_0(\sigma) = \phi_0(\sigma) - \int \phi_0(\sigma)\mu(d\sigma)$. Then $R(t) = \int T_t\bar{\phi}_0(\sigma)\bar{\phi}_0(\sigma)\mu(d\sigma)$.

Let $E_\lambda$ be the spectral resolution of the identity corresponding to $-\mathcal{L}$. Then

$$\int T_t\bar{\phi}_0(\sigma)\bar{\phi}_0(\sigma)\mu(d\sigma) = \int_0^\infty e^{-\lambda t} d_\lambda(E_\lambda\bar{\phi}_0, \bar{\phi}_0).$$

If we let $\epsilon = \inf\{\lambda : (E_\lambda\bar{\phi}_0, \bar{\phi}_0) > 0\}$, then for all $\delta > 0$

$$\int_0^\infty e^{-\lambda t} d_\lambda(E_\lambda\bar{\phi}_0, \bar{\phi}_0) \geq e^{-(\epsilon+\delta)t}(E_{\epsilon+\delta}\bar{\phi}_0, \bar{\phi}_0)$$

and

$$\int_0^\infty e^{-\lambda t} d_\lambda(E_\lambda\bar{\phi}_0, \bar{\phi}_0) \leq e^{-\epsilon t}\|\bar{\phi}_0\|^2.$$

Therefore for all $t > 0$

$$\frac{-1}{t} \log(R(t)) \leq (\epsilon + \delta) - \frac{1}{t} \log((E_{\epsilon+\delta}\bar{\phi}_0, \bar{\phi}_0)),$$

and

$$\frac{-1}{t}\log(R(t)) \geq \epsilon - \frac{1}{t}\|\bar{\phi}_0\|^2.$$

Thus

$$\overline{\lim_{t\to\infty}} \frac{-1}{t}\log(R(t)) \leq (\epsilon+\delta) \qquad \text{for all } \delta > 0,$$

and

$$\underline{\lim_{t\to\infty}} \frac{-1}{t}\log(R(t)) \geq \epsilon.$$

Q.E.D.

We denote $\lim_{t\to\infty} \frac{-1}{t}\log(R(t)) = \epsilon$. A close look at the proof of Theorem (2.1) reveals that $\epsilon$ is at least as big as the gap between 0 and the rest of the spectrum of $-\mathcal{L}$. Our next goal is to prove the opposite inequality.

First we define

$$gap = \inf\{\frac{-\int(f,\mathcal{L}f)d\mu}{\|f\|^2_{L^2(\mu)}} : f \in L^2(\mu) \text{ and } \int f d\mu = 0\}.$$

**Theorem (2.3).** *For any stochastic* Ising *model with ferromagnetic potential and attractive interactions,*

(2.4) $$\lim_{t\to\infty} \frac{-1}{t}\log(R(t)) = gap.$$

Before proving this theorem we prepare some lemmas.

**Lemma (2.5).** *Let $\sigma(t)$ be a stochastic* Ising *model with ferro-magnetic potential and attractive flip rates. Define*

$$\chi_A(\sigma) = \prod_{x\in A}\frac{1+\sigma(x)}{2} - \int\prod_{x\in A}\frac{1+\sigma(x)}{2}d\mu(\sigma).$$

*Then*

(2.6) $$\underline{\lim_{t\to\infty}} \frac{-1}{t}\log(\int T_t\chi_A(\sigma)\chi_A(\sigma)\mu(d\sigma)) \geq \epsilon.$$

*Proof.* Set $\bar{\phi}_k(\sigma) = \phi_k(\sigma) - \int \phi_k(\eta)\mu(d\eta)$, and then let $\psi_A(\sigma) = \sum_{x\in A} \bar{\phi}_x(\sigma) - \chi_A(\sigma)$. It is easily seen that $\psi_A(\sigma)$ is an increasing function of $\sigma$ and thus, since the interaction is attractive, $T_t\psi(\sigma)$ is also an increasing function of $\sigma$. From the F.K.G. inequality (see [FKG]) we therefore have

$$\int \chi_A(\sigma)T_t\psi_A(\sigma)\mu(\sigma) \geq \int \chi_A(\sigma)\mu(d\sigma)\int T_t\psi_A(\sigma)\mu(d\sigma) = 0.$$

Equivalently

$$\sum_{x\in A}\int \chi_A(\sigma)T_t\bar{\phi}_x(\sigma)\mu(d\sigma) \geq \int \chi_A(\sigma)T_t\chi_A(\sigma)\mu(d\sigma).$$

Thus

$$\begin{aligned}\int \chi_A(\sigma)T_t\chi_A(\sigma)d\mu &\leq \sum_{x\in A}\|\chi_A\|_{L^2(\mu)}\|T_t\bar{\phi}_0\|_{L^2(\mu)}\\ &=| A | \, \|\chi_A\|_{L^2(\mu)}\left(\int \bar{\phi}_0 T_{2t}\bar{\phi}_0 d\mu\right)^{1/2}.\end{aligned}$$

Finally we have

$$\begin{aligned}&\underline{\lim}_{t\to\infty}\frac{-1}{t}\log\left(\int \chi_A(\sigma)T_t\chi_A(\sigma)\mu(d\sigma)\right)\\ &\geq \underline{\lim}_{t\to\infty}\frac{-1}{t}\left[\log(| A |) + \log(\|\chi_A\|_{L^2(\mu)}) + \frac{1}{2}\log\left(\int \bar{\phi}_0 T_{2t}\bar{\phi}_0 d\mu\right)\right]\\ &= \epsilon.\end{aligned}$$

Q.E.D.

**Lemma (2.7).** *Under the hypotheses of Lemma* (2.5), *if $f$ is a cylinder function with $\int f d\mu = 0$ then*

$$\underline{\lim}_{t\to\infty}\frac{-1}{t}\log\left(\int T_t f(\sigma)f(\sigma)\mu(d\sigma)\right) \geq \epsilon. \tag{2.8}$$

*Proof.* Every cylinder function $f$ with $\int f d\mu = 0$ may be written as

$$f(\sigma) = \sum_A \alpha_A \chi_A(\sigma),$$

where all $A \neq \emptyset$ and only finitely many $\alpha_A \neq 0$. Therefore,

$$\begin{aligned}
&\int f(\sigma) T_t f(\sigma) \mu(d\sigma) \\
&\quad = \int \sum_A \alpha_A \chi_A(\sigma) T_t (\sum_B \alpha_B \chi_B)(\sigma) \mu(d\sigma) \\
&\quad \leq \sum_A \sum_B | \alpha_A || \alpha_B | \int \chi_A(\sigma) T_t \chi_B(\sigma) \mu(d\sigma) \\
(2.9) \quad &\quad \leq \sum_A \sum_B | \alpha_A || \alpha_B | \, \|\chi_A\|_{L^2(\mu)} \|T_t \chi_B\|_{L^2(\mu)} \\
&\quad = \sum_A \sum_B | \alpha_A || \alpha_B | \, \|\chi_A\|_{L^2(\mu)} \left( \int \chi_B(\sigma) T_{2t} \chi_B(\sigma) \mu(d\sigma) \right)^{1/2} \\
&\quad \leq \sum_A \sum_B | \alpha_A || \alpha_B | \, \|\chi_A\|_{L^2(\mu)} \max_{C:\alpha_C \neq 0} \left( \int \chi_C(\sigma) T_{2t} \chi_C(\sigma) \mu(d\sigma) \right)^{1/2}.
\end{aligned}$$

Taking the logarithm of both sides of (2.9) and dividing by $t$ yields

$$\begin{aligned}
&\varliminf_{t\to\infty} \frac{-1}{t} \log \left( \int f(\sigma) T_t f(\sigma) \mu(d\sigma) \right) \\
&\quad \geq \varliminf_{t\to\infty} \frac{-1}{t} \left( \log \left( \sum_A \sum_B | \alpha_A || \alpha_B | \, \|\chi_A\|_{L^2(\mu)} \right) \right. \\
&\qquad\qquad \left. + \max_{B:\alpha_B \neq 0} \frac{1}{2} \log \left( \int \chi_B T_{2t} \chi_B d\mu \right) \right) \\
&\quad = \epsilon.
\end{aligned}$$

Q.E.D.

*Proof of Theorem* (2.3). By the remark following Lemma (2.1) $gap \leq \epsilon$. Suppose that $gap < \epsilon$. Then there exists function $f \in L^2(\mu)$ and a $\delta > 0$ such that $E_{(0,\epsilon-\delta)}(f) = f$ and $\|f\|_{L^2(\mu)} = 1$. Since the cylinder functions are dense in $L^2$ there is a cylinder function $g$ such that $\|f-g\| < \frac{1}{2}$; and since $\int f d\mu = 0$ we may assume that $\int g d\mu = 0$. Therefore $\|E_{(0,\epsilon-\delta)}(g-f)\| \leq \|g-f\| \leq \frac{1}{2}$. It follows that $\|E_{(0,\epsilon-\delta)}(g)\| = \|E_{(0,\epsilon-\delta)}(f) + E_{(0,\epsilon-\delta)}(g-f)\| \geq \|E_{(\epsilon-\delta)}(f)\| - \|E_{(0,\epsilon-\delta)}(g-f)\| > \frac{1}{2}$. Thus

$$\begin{aligned} \int g(\sigma)T_t g(\sigma)\mu(d\sigma) &= \int_0^\infty e^{-\lambda t} d_\lambda(E_\lambda g, g) \\ &\geq \int_0^{\epsilon-\delta} e^{-\lambda t} d_\lambda(E_\lambda g, g) \\ &\geq e^{-t(\epsilon-\delta)}\|E_{(0,\epsilon-\delta)}g\|^2. \end{aligned} \tag{2.10}$$

From (2.10) we obtain

$$\overline{\lim_{t\to\infty}} \frac{-1}{t} \log\left(\int g(\sigma)T_t g(\sigma)\mu(d\sigma)\right) \leq \epsilon - \delta.$$

This contradicts Lemma (2.7). Q.E.D.

We have now proved that the exponential rate at which the autocorrelation function of a single spin decays to zero is the same as the gap in the spectrum of the generator. Theorem (2.11) below gives an upper bound on this gap. The bound in Theorem (2.11) is well known, but since the proof is easy we include it for completeness. The bound is given in terms of the, susceptibility, $\chi$, which is defined by

$$\chi = \sum_{x\in Z^d} \int \bar{\phi}_0(\sigma)\bar{\phi}_x(\sigma)\mu(\sigma).$$

**Theorem (2.11).** $gap \leq \dfrac{2\int c_0(\sigma)\mu(d\sigma)}{\chi}$.

*Proof.* In view of the definition of *gap*, for any cylinder function $f$ with $\inf f d\mu = 0$ we have

$$gap \leq \frac{-\int f\mathcal{L}f d\mu}{\|f\|^2_{L^2(\mu)}}. \tag{2.12}$$

In (2.12) we take $f$ to be $f_n(\sigma) = \sum_{|x|<n} \bar{\phi}_x(\sigma)$. Note that

$$-\int f_n(\sigma)\mathcal{L}f_n(\sigma) = \frac{1}{2}\sum_{|x|<n}\int (f_n(\sigma^x) - f_n(\sigma))^2 c_x(\sigma)\mu(d\sigma)$$

and

$$(f_n(\sigma^x) - f_n(\sigma))^2 \equiv 4$$

for all $x \in \{y :| y |< n\}$ and all $\sigma \in \mathbf{E}$. Therefore

$$-\int f_n(\sigma)\mathcal{L}f_n(\sigma)\mu(d\sigma) = 2\#\{y < n\}\int c_0(\sigma)\mu(d\sigma).$$

Also

$$\|f_n\|^2_{L^2(\mu)} = \sum_{|x|<n}\sum_{|y|<n}\int \bar{\phi}_x(\sigma)\bar{\phi}_y(\sigma)\mu(d\sigma).$$

Combining these two observations we have

$$gap \leq \frac{2\#\{y<n\}\int c_0(\sigma)\mu(d\sigma)}{\sum_{|x|<n}\sum_{|y|<n}\int \bar{\phi}_x(\sigma)\bar{\phi}_y(\sigma)\mu(d\sigma)}.$$

Letting $n$ go to $\infty$ we obtain the desired result. Q.E.D.

It is very difficult to compute $\chi$ for any potential if $d > 1$; however, the asymptotics of $\chi$ as the temperature decreases to the critical temperature have been determined in the case when $d = 2$ and the potential satisfies $J_{\{x,y\}} = \frac{1}{T}$ if $| x - y |= 1$ and 0 in all other cases. In this case $\chi(T) \approx| T - T_c |^{-7/4}$. Thus in this case *gap* as a function of temperature goes to zero at least as fast as $| T - T_c |^{7/4}$ .

## §3 Rate of convergence at the critical temperature

In this section we first mention a general result about how fast the semigroup evaluated on cylinder functions can converge when the system is exactly at the critical temperature. We then give a considerable improvement on this general rate in the case of a two dimensional stochastic ISING model with nearest neighbor ferromagnetic pair potentials.

Our general result relies in the following theorem, the proof of which may be found in [H].

**Theorem (3.1).** *Consider the set of spin flip processes whose flip rates are attractive, translation invariant, and have a range of interaction which is bounded by some fixed constant. Each process in such a set may be identified with its flip rate at 0. The subset of those processes which converge exponentially fast to their equilibrium in the sense of $L^\infty$ convergence of the semi-group evaluated on cylinder functions is open in the topology of uniform convergence of the flip rates.*

**Theorem (3.2).** *Let $\{T_t : t \geq 0\}$ be the semi-group for a stochastic* ISING *model with finite range attractive flip rates and a potential which is exactly at the critical temperature. Then there is a constant $c > 0$ such that for all $t \geq 1$*

$$T_t\phi_0(\vec{1}) - T_t\phi_0(\overrightarrow{-1}) \geq ct^{-d}. \tag{3.3}$$

*Proof.* It is proved in [A&H2] and in [H] that there are only two possibilities. Either (3.3) holds, or the expression on the left side of (3.3) converges to zero exponentially fast. However, if the latter alternative holds then by Theorem (3.1) it also holds at all near by potentials. Since we are exactly at the critical temperature there are arbitrarily close potentials for which there are two GIBBS states, and thus for which the quantity on the left side of (3.3) does not even go to zero. Thus the only possibility is for (3.3) to hold.Q.E.D.

For the rest of this paper we specialize to the case of $d = 2$ and $J_{\{x,y\}} = \beta_c$ if $| x - y |= 1$ and 0 otherwise. Here $\beta_c = \frac{1}{2}\log(1+\sqrt{2})$, the critical value for the two dimensional nearest neighbor pair potential ISING model. For this model the following result is known (see [M&W]).

**Lemma (3.4).** *Let $\vec{k} = (k, 0)$. Then*

$$\lim_{k\to\infty} -\frac{\log(\int \phi_{\vec{0}}(\sigma)\phi_{\vec{k}}(\sigma)\mu(d\sigma))}{\log(k)} = \frac{1}{4}. \tag{3.5}$$

**Theorem (3.6).** *For the nearest neighbor stochastic* ISING *model with ferromagnetic pair potentials exactly at the critical temperature we have*

$$\overline{\lim_{t\to\infty}} -\frac{\log(R(t))}{\log(t)} \leq \frac{1}{4}. \tag{3.7}$$

*Proof.* We first observe that by [H&S, Remark (2.21)] $R(t)$ must go to 0 more slowly than exponentially as $t \to \infty$. Also by [H&S, Theorem (2.11)] there are constants $\alpha < \infty$ and $\gamma > 0$ such that for all $k \geq 0$

$$\sup_{0 \leq t \leq k/\alpha} | T_t(\phi_{\vec{0}}(\cdot)\phi_{\vec{k}}(\cdot))(\sigma) - T_t\phi_{\vec{0}}(\sigma)T_t\phi_{\vec{k}}(\sigma) | = O(e^{-\gamma k}). \tag{3.8}$$

Let $k = [\alpha t + 1]$. Then since for any $k$ and any $t \geq 0$,

$$\int \phi_{\vec{0}}(\sigma)\phi_{\vec{k}}(\sigma)\mu(d\sigma) = \int T_t(\phi_{\vec{0}}(\cdot)\phi_{\vec{k}}(\cdot))(\sigma)\mu(d\sigma),$$

we have

$$\begin{aligned}
&\int \phi_{\vec{0}}(\sigma)\phi_{\vec{k}}(\sigma)\mu(d\sigma) \\
&\quad = \int T_t\phi_{\vec{0}}(\sigma)T_t\phi_{\vec{k}}(\sigma)\mu(d\sigma) + O(e^{-\gamma k}) \\
&\quad \leq \sqrt{\int (T_t\phi_{\vec{0}}(\sigma))^2\mu(d\sigma) \int (T_t\phi_{\vec{k}}(\sigma))^2\mu(d\sigma)} + O(e^{-\gamma k}) \\
&\quad = \int T_t\phi_{\vec{0}}(\sigma)T_t\phi_{\vec{0}}(\sigma)\mu(d\sigma) + O(e^{-\gamma k}) \\
&\quad = \int \phi_{\vec{0}}(\sigma)T_{2t}\phi_{\vec{0}}(\sigma)\mu(d\sigma) + O(e^{-\gamma k}).
\end{aligned}$$

Now recalling Lemma (3.4) and that $k = [\alpha t + 1]$ we have

$$\begin{aligned}
\frac{1}{4} &= \lim_{t \to \infty} -\frac{\log(\int \phi_{\vec{0}}(\sigma)\phi_{\vec{k}}(\sigma)\mu(d\sigma))}{\log([\alpha t + 1])} \\
&\geq \overline{\lim_{t \to \infty}} -\frac{\log(\int \phi_{\vec{0}}(\sigma)T_{2t}\phi_{\vec{0}}(\sigma)\mu(d\sigma) + O(e^{-\alpha\gamma t}))}{\log([\alpha t + 1])} \\
&= \overline{\lim_{t \to \infty}} \left( -\frac{\log(2t)}{\log([\alpha t + 1])} \left( \frac{\log(R(2t))}{\log(2t)} + \frac{\log(1 + \frac{O(e^{-\alpha\gamma t})}{R(2t)})}{\log(2t)} \right) \right).
\end{aligned} \tag{3.9}$$

By the remark at the beginning of the proof the second term on the right side of (3.9) goes to zero and hence we conclude from (3.9) that

$$\frac{1}{4} \geq \overline{\lim_{t\to\infty}} -\frac{\log(R(t))}{\log(t)}.$$

Q.E.D.

*Remark (3.10).* Note that

$$R(t) \leq \|\phi_{\vec{0}}\|_{L^2(\mu)} \|T_t\phi_{\vec{0}}\|_{L^2(\mu)} \leq \|T_t\phi_{\vec{0}}\|_{L^\infty} = T_t\phi_{\vec{0}}(\vec{1}\,) \vee (-T_t\phi_{\vec{0}}(\overrightarrow{-1})).$$

Thus

$$\underline{\lim_{t\to\infty}} -\frac{\log(T_t\phi_{\vec{0}}(\vec{1}\,) - T_t\phi_{\vec{0}}(\overrightarrow{-1}))}{\log(t)} \leq \frac{1}{4}.$$

## References


[A&H1] R. Abe and A. Hatano, *Dynamics of the Ising Model Near the Transition Point*, Progress of Theoretical Physics **39** (1968), pp. 947-956.

[A&H2] M. Aizenman and R. Holley, *Rapid Convergence to Equilibrium of Stochastic Ising Models in the Dobrushin Shlosman Regime*, "Percolation Theory and Ergodic Theory of Infinite Particle Systems," IMA Vols in Math. and its Appl. ed by Harry Kesten, Springer Verlag, 1987, pp. 1-11.

[B] K. Binder, *Monte Carlo Investigations of Phase Transitions and Critical phenomena*, Phase Transitions and Critical Phenomena **5B** (1976), pp. 1-105.

[B&K] K. Binder and M.H. Kalos, *Monte Carlo Studies of Relaxation Phenomena: Kinetics of Phase Changes and Critical Slowing Down*, "Monte Carlo Methods in Statistical Physics," 1979, pp. 225-260.

[FKG] C.M. Fortuin, P.W. Kasteleyn, and J. Ginibre, *Correlation Inequalities on Some Partially Ordered Sets*, Commun. Math. Phys. **22** (1971), pp. 89-103.

[H] R. Holley, *Possible Rates of Convergence in Finite Range, Attractive Spin Systems*, Contemporary Math. **41** (1985), pp. 215-234..

[H&H] B.I. Helpern, and P.C. Hohenberg, *Theory of Dynamic Critical Phenomena*, Rev. Mod. Phys. **49** (1977), pp. 435-479..

[H&S] R. Holley and D. W. Stroock, *$L_2$ theory for the stochastic Ising model*, Z. Wahr. verw. Geb. **35** (1976), pp. 87-101.

[K] K. Kawasaki, *Kinetics of Ising Models*, Phase Transitions and Critical Phenomena **2** (1972), pp. 443-501.

[L] T. M. Liggett, "Interacting Particle Systems," Springer Verlag, 1985.

[M&W] B.M. McCoy and T.T Wu, "The Two-Dimensional Ising Model," Harvard University Press, Cambridge, Massachusets, 1973.


This work was partially supported by N.S.F. Grant DMS-8609944.

Department of Mathematics
University of Colorado
Boulder, CO 80309

# Spatially Inhomogeneous Contact Processes

Thomas M. Liggett

## Summary

A one-dimensional spatially inhomogeneous contact process is a Markov process in $\{0,1\}^Z$, where $Z$ is the set of integers, which has the following transitions:

$$1 \to 0 \text{ at site } k \text{ at rate } \delta(k), \text{ and}$$

$$0 \to 1 \text{ at site } k \text{ at rate } \rho(k)\eta(k+1) + \lambda(k)\eta(k-1),$$

where $\delta(k) > 0, \rho(k) \geq 0, \lambda(k) \geq 0$, and $\eta \in \{0,1\}^Z$ is the current configuration. If $\delta(k) \equiv 1$ and $\rho(k) \equiv \lambda(k) \equiv \lambda$, this is the basic contact process which was first studied by Harris in 1974. If $\{(\delta(k), \rho(k), \lambda(k)), k \in Z\}$ is chosen randomly in a stationary ergodic manner, it is natural to call this a contact process in a random environment. In this paper, we present three types of results, giving sufficient conditions (a) for extinction of the process, (b) for survival of the process, and (c) for the process to have at most four extremal invariant measures.

## 1. Introduction.

The basic contact process plays much the same central role in the area of interacting particle systems that Brownian motion plays in the theory of stochastic processes. It is a simple, easily defined process which has a rich, nontrivial theory. It has most of the important properties which are useful in the analysis of an interacting particle system. It is attractive, additive and self-dual. It has nearest neighbor interactions, and it is interesting even in one dimension. It does fail to have two useful properties which some other particle systems enjoy. It is not reversible, as are the stochastic Ising models and some nearest particle systems for example, and its dual is not bounded as is the dual of the voter model. However, it is the absence of these properties which makes the behavior and analysis of the contact process interesting and nontrivial. An indication of the importance of the contact process is the fact that it is the subject of over 25 papers, and of chapters in the three monographs on particle systems: Griffeath (1979), Liggett (1985), and Durrett (1988). The latter two monographs and the survey paper by Griffeath (1981) can be consulted for detailed information about this process.

The contact process was introduced and first studied by Ted Harris fifteen years ago. In his 1974 paper, he proved that the critical value for the basic contact process in $d$ dimensions is finite, and is at least $(2d-1)^{-1}$. In one dimension, he obtained the better lower bound of 1.18. In the same paper, he studied and exploited various monotonicity and convexity properties in this context. In three succeeding papers, Harris (1976, 1977, 1978), he developed a number of tools which have turned out to be essential in the analysis of large classes of interacting particle systems which contain the contact process as a special case. These include duality (which he called "association"), the preservation in time of positive correlations of the distributions of certain Markov processes, and perhaps most importantly, the use of graphical or percolation techniques. In this paper, we will consider a spatially inhomogeneous version of the one dimensional contact process. By this we mean a Markov process on $\{0,1\}^Z$,where $Z$ is the set of integers, which has the following transitions:

$$1 \to 0 \text{ at site } k \text{ at rate } \delta(k), \text{ and}$$

$$0 \to 1 \text{ at site } k \text{ at rate } \rho(k)\eta(k+1) + \lambda(k)\eta(k-1),$$

where $\delta(k) > 0, \rho(k) \geq 0, \lambda(k) \geq 0$, and $\eta \in \{0,1\}^Z$ is the current configuration. If $\delta(k) \equiv 1$ and $\rho(k) \equiv \lambda(k) \equiv \lambda$, this is the basic contact process. If $\{(\delta(k), \rho(k), \lambda(k)), k \in Z\}$ is chosen randomly in a stationary ergodic manner, we will call this a contact process in a random environment. Interacting particle systems in random environments have been considered in a number of papers, including Greven (1985), Ferreira (1988), and Bramson, Durrett and Schonmann (1989). Two graduate students at UCLA, D. Chen and X. Liu, are currently working in reversible nearest particle systems in random environments. Random walks and diffusions in random environments have been treated in several papers, including Solomon (1975), Kesten, Kozlov and Spitzer (1975), Sinai (1982), Brox (1986), and Durrett (1986).

In order to apply standard existence and uniqueness theorems, such as Theorem I.3.9 in Liggett (1985), to the spatially inhomogeneous contact process, one would need to assume that the sequences $\delta(k), \rho(k)$, and $\lambda(k)$ are uniformly bounded. Little of interest is lost by making that assumption. Alternatively, one can give a simple direct argument based on coupling which shows that the process is well defined provided that

$$\sum_{k\geq n}\frac{1}{\rho(k)}=\infty \text{ and } \sum_{k\leq n}\frac{1}{\lambda(k)}=\infty \tag{1.1}$$

for all $n$. This condition is automatically satisfied a.s. in the random environment case. Thus we will not dwell on technical issues related to the construction of the process, but will always assume at least that (1.1) holds.

The spatially inhomogeneous contact process retains many of the characteristics of the basic contact process. It is attractive, and it has a dual which is another process of the same type. Specifically, if $\eta_t$ is the process with rates $\delta(k), \rho(k)$, and $\lambda(k), \zeta_t$ is the process with rates $\delta'(k), \rho'(k)$, and $\lambda'(k)$ and $A_t = \{k : \zeta_t(k) = 1\}$, then $\eta_t$ and $\zeta_t$ satisfy the duality equation

$$p^{\eta}\left[\eta_t(k)=0 \text{ for all } k\in A\right]=P^{A}\left[\eta(k)=0 \text{ for all } k\in A_t\right] \tag{1.2}$$

for finite $A$ provided that $\delta'(k) = \delta(k), \rho'(k) = \lambda(k+1)$ and $\lambda'(k) = \rho(k-1)$. Note that (1.1) implies that $A_t$ is finite a.s. if $A_0$ is finite, so that (1.2) could be used to define the process $\eta_t$.

We will say that the infinite system survives if it has an invariant measure other than the pointmass on the identically zero configuration, and that the finite system survives if $A_t$ is nonempty for all $t$ with positive probability for some finite initial set $A_0$. By (1.2), the infinite system survives if and only if the dual finite system survives. Since survival is preserved by the operator of reflection in $Z$, it follows that the two concepts of survival coincide in the random environment case, provided that the birth an death rates are mutually independent, with distributions which are independent of $k$. The two concepts do not coincide for general inhomogeneous contact processes. For example, the infinite system survives and the finite system dies out if $\delta(k) = 1$ for all $k, \lambda(k) \geq 4$ and $\rho(k) \leq 1-\varepsilon$ for $k < 0$ and $\rho(k) \geq 4$ and $\lambda(k) \leq 1-\varepsilon$ for $k > 0$, where $\varepsilon > 0$. (To check the survival of the infinite system, compare it with a one-sided homogeneous contact process, and apply Theorem 1.3 of Holley and Liggett (1978). To see that the finite system dies out, apply Theorem 1 below to the dual process). The dual of such a system has the property that the finite system survives and the infinite system dies out.

Our main purpose here is to try to determine when the infinite system survives. Of course, a comparison with the basic contact process shows that the system survives if all the $\lambda(k)$'s and $\rho(k)$'s are substantially larger than all the $\delta(k)$'s, while the system dies out if all the $\lambda(k)$'s and $\rho(k)$'s are substantially smaller than all the $\delta(k)$'s. The point is to obtain conditions for survival and extinction which allow some of the $\lambda(k)$'s, $\rho(k)$'s and $\delta(k)$'s to be large and some to be small. As in the case of the basic contact process, it is not hard to give sufficient conditions for extinction. The first four theorems give various such conditions, and will be proved in the next section.

**Theorem 1.** *The infinite spatially inhomogeneous contact process dies out provided*

$$\sum_{k\geq n}\prod_{j=n}^{k}\frac{\rho(j)}{\delta(j+1)}<\infty \quad \text{and} \quad \sum_{k\leq n}\prod_{j=k}^{n}\frac{\lambda(j)}{\delta(j-1)}<\infty \tag{1.3}$$

*for all n.*

**Corollary 1.** *The infinite contact process in a random environment dies out a.s. provided that*

$$E\log\rho(0)<E\log\delta(0) \quad \text{and} \quad E\log\lambda(0)<E\log\delta(0) \tag{1.4}$$

*where the expected values above are assumed to be well defined.*

*Proof.* By the ergodic theorem,

$$\lim_{k\to\infty}\frac{1}{k}\sum_{j=0}^{k}\log\frac{\rho(j)}{\delta(j+1)}=E\log\frac{\rho(0)}{\delta(0)} \text{ a.s.}$$

and

$$\lim_{k\to\infty}\frac{1}{k}\sum_{j=-k}^{0}\log\frac{\lambda(j)}{\delta(j-1)}=E\log\frac{\lambda(0)}{\delta(0)} \text{ a.s.}$$

These limits are strictly negative by (1.4). Therefore (1.3) is satisfied by almost every environment.

One might be tempted to guess from Corollary 1 that a contact process in a random environment with $\delta(k)\equiv 1$ survives if either $E\log\rho(0)$ or $E\log\lambda(0)$

is sufficiently large. The point of the next theorem and its corollary is to show that this is not the case, at least for one-sided systems. It remains an open problem to determine whether this guess is correct for systems which have sufficient symmetry - for example if the $\lambda$'s and $\rho$'s have the same distribution. See Example 3.25 for some computer calculations which suggest conjectures along these lines. The proof of Theorem 2 follows the proof of Theorem 7.1 in Harris (1974), which gives the main lower bound for the critical value of the basic contact process. Corollary 2 follows Theorem 2 in the same way that Corollary 1 followed from Theorem 1.

**Theorem 2.** *The infinite spatially inhomogeneous contact process with* $\delta(k) \equiv 1$ *and* $\lambda(k) \equiv 0$ *dies out provided that*

$$\prod_{k \geq n} \frac{2\rho(k)}{2+\rho(k)} = 0$$

*for all* $n$.

**Corollary 2.** *The infinite contact process in a random environment with* $\delta(k) \equiv 1$ *and* $\lambda(k) \equiv 0$ *dies out provided that*

$$E \log \frac{2\rho(k)}{2+\rho(k)} < 0.$$

Next, we have a sufficient condition for extinction when the environment is i.i.d., but is not one-sided as in the case of Corollary 2, nor essentially symmetric in $\lambda$ and $\rho$ as in the case of Corollary 1.

**Theorem 3.** *Suppose that the random vectors* $\{(\delta(k), \rho(k), \lambda(k)), k \in Z\}$ *are i.i.d. The infinite contact process in a random environment dies out if*

$$E\rho(k)\,[\delta(k)]^{-1} < 1 \qquad \text{and}$$

$$E\,[\delta(k)]^{-1}\left\{1 - E\delta(k)\,[\lambda(k)]^{-1}\right\} < E\,[\lambda(k)]^{-1}\left\{1 - E\rho(k)\,[\delta(k)]^{-1}\right\},$$

*where all the expectations are assumed to be finite.*

*Remarks.*

(a) The proof of Theorem 3 is based on the theory of random walks in random environments - specifically on continuous time versions of results in Solomon (1975). Other sufficient conditions for extinction could be obtained from continuous time versions of the results in Kesten, Kozlov and Spitzer (1975).
(b) If the $\rho(k)$'s in Corollary 2 are i.i.d, then that result follows from Theorem 3 and Jensen's inequality.
(c) In Theorem 3, the roles of $\rho$ and $\lambda$ can of course be interchanged.
(d) Consider the example in which $\delta(k) \equiv 1$, the $\rho(k)$'s are i.i.d. uniform on $[0, R]$ and the $\lambda(k)$'s are i.i.d uniform on $[0, L]$. In this case, Corollary 1 gives extinction if $R < e$ and $L < e$, while Theorem 3 gives extinction if $R < 2$ or $L < 2$.

A natural question is whether a contact process in a random environment is more or less likely to survive than the contact process whose birth and death rates are obtained from those of the original by taking expectations. The next result gives an answer to this question in a one-sided situation. Its proof uses the fact which was first proved in Harris (1960) that increasing functions of independent random variables are positively correlated.

**Theorem 4.** *Consider the infinite contact process in a random environment $\eta_t$ with $\lambda(k) \equiv 0, \delta(k) \equiv 1$, and $\rho(k)$ independent and identically distributed, and let $\zeta_t$ be the process with deterministic environment $\lambda'(k) \equiv 0, \delta'(k) \equiv 1$ and $\rho'(k) \equiv E\rho(k)$. If $\zeta_t$ dies out, then so does $\eta_t$ for almost every environment.*

*Remarks.*

(a) It follows from Corollary 1 and Jensen's inequality that the $\eta_t$ in Theorem 4 dies out if $E\rho(k) < 1$. A similar argument based on Corollary 2 shows that it dies out if $E\rho(k) < 2$. Since the critical value if the one-sided homogeneous contact process is (substantially) greater than two, Theorem 4 is not a consequence of either of these corollaries.
(b) It would be interesting to obtain analogues of Theorem 4 in the two-sided case, and for similar models in higher dimensions. In particular, such a result in higher dimensions would give a useful sufficient condition for extinction of multidimensional contact processes in i.i.d.

random environments. Holley has pointed out that no analogue of Corollary 1 can hold in dimensions greater than one. To see this, take $\delta(k) \equiv 1$ and the birth rate at $k$ to be of the form

$$\sum_j \lambda(k,j)\,\eta(j),$$

where the sum is over lattice points in $Z^d$ which neighbor $k$, and the $\lambda$'s are assigned in an i.i.d. fashion to the (undirected) edges on $Z^d$. If $P[\lambda(k,j) \geq 2]$ is greater than the critical value for ordinary bond percolation on $Z^d$, then with probability one, there will be an embedded one dimensional supercritical basic contact process, and hence the process in the random environment will survive. The assumption that $E\lambda(k,j)$ is small rules out that possibility, but the assumption that $E\log\lambda(k,j)$ is small does not.

Finding good conditions for the survival of spatially inhomogeneous contact processes is much more difficult. The best sufficient condition for survival of the basic contact process is $\lambda \geq 2$. As it stands, the proof of the result which is given in Holley and Liggett (1978) cannot be applied to spatially inhomogeneous processes. However, we will now describe an improvement of that approach which does yield a sufficient condition for the survival of spatially inhomogeneous contact processes. It is easy to check this condition in the case of the basic contact process with $\lambda \geq 2$. Checking it in greater generality is more difficult, but it can be done in some cases.

The Holley-Liggett proof that the basic contact process survives consists of the following two steps:

(1) Show that the contact process survives if there is a logarithmically convex probability density $f(n)$ on the positive integers with finite mean which satisfies the following system of nonlinear equations:

$$(2+2\lambda)\,f(n) = 2\lambda f(n+1) + \sum_{k=1}^{n-1} f(k)\,f(n-k) \quad \text{for } n \geq 2 \tag{1.5}$$

(2) Show that such a solution to (1.5) exists if (and only if) $\lambda \geq 2$.

The second step was quite easy, since (1.5) is a convolution equation, The hard work came in carrying out the first step. Note, however, that the magnitude of $\lambda$ is only relevant in the second step.

The following result, which is proved in Section 3, gives an analogue if the first step above, which applies to all spatially inhomogeneous contact processes. Unfortunately, the second step now becomes substantially more difficult, we will have to leave as an open problem the determination of good conditions on the transition rates which guarantee that the assumptions of Theorem 5 below are satisfied. One example will be worked out explicitly in Section 3 as Example 3.20: if $\lambda(2n+1) = \rho(2n) = \lambda_1$, $\rho(2n+1) = \lambda(2n) = \lambda_2$ and $\delta(n) = \delta$, then the conditions of Theorem 5 are satisfied provided that $\lambda_1\lambda_2 \geq 4\delta^2$. By Theorem 1, this system dies out if $\lambda_1\lambda_2 < \delta^2$. Examples 3.24 and 3.25 give numerical results which indicate in some other cases how generally one should expect Theorem 5 to apply.

**Theorem 5.** *Suppose that there are functions $\alpha(i) > 0$ and $\phi(k,l) \geq 0$, $-\infty < k < l < \infty$, which satisfy*

$$\sum_{k\leq i<l} \phi(k,l) = 1,$$

$$\alpha(i) = \sum_{k>i} \phi(i,k) = \sum_{k<i} \phi(k,i)$$

*and*

$$0 = \lambda(i)\phi(i-1,i+1) + \rho(i+1)\phi(i,i+2) + \lambda(i+1)\alpha(i) + \rho(i)\alpha(i+1)$$
$$- [\delta(i) + \delta(i+1) + \rho(i) + \lambda(i+1)]\phi(i,i+1)$$

*for all $i$, and*

$$0 = \lambda(k)\phi(k-1,l) + \rho(l)\phi(k,l+1) + \sum_{j=k+1}^{l-1} \frac{\delta(j)}{\alpha(j)}\phi(k,j)\phi(j,l)$$
$$- [\delta(k) + \delta(l) + \lambda(k+1) + \rho(l-1)]\phi(k,l)$$

*and*

$$\phi(k,l-1)\,\phi(k-1,l) \geq \phi(k,l)\,\phi(k-1,l-1)$$

*for* $k < l-1$. *Then the infinite contact process survives.*

*Remark.* To make the connection between the assumption in Theorem 5 and equation (1.5), take $\delta(i) \equiv 1, \lambda(i) \equiv \lambda, \rho(i) \equiv \lambda, \alpha(i) \equiv \alpha =$ the reciprocal of the mean of the density $f(n)$, and $\phi(k,l) = \alpha f(l-k)$.

Our final result describes the totality of all extremal invariant measures for the inhomogeneous contact process. Under mild assumptions, it states that the process can have 1, 2, 3 or 4 extremal invariant measures. Let $v_0$ be the pointmass on $\eta \equiv 0$ and $v_1$ be the upper invariant measure of the contact process. The proof of the following theorem is an adaptation of the proof of Theorem 3.13 of Chapter III of Liggett (1985), will be given in Section 4.

*Remark.* Suppose that the $\lambda$'s $\rho$'s, and $\delta$'s are uniformly bounded, and that

$$\inf \delta(j)\,(|j|+1)^{\varepsilon} > 0 \text{ for every } \varepsilon > 0. \tag{1.6}$$

If $v$ is an extremal invariant measure for the inhomogeneous contact process which is different from $v_0$ and $v_1$, then it can be coupled with $v_1$ in such a way that the coupling measure concentrates (i) on

$$\{(\eta,\zeta):\ \text{there is a } k \text{ so that } \eta \equiv 0 \text{ on } (-\infty,k) \text{ and } \eta \equiv \zeta \text{ on } [k,\infty)\}$$

or (ii) on

$$\{(\eta,\zeta):\ \text{there is a } k \text{ so that } \eta \equiv \zeta \text{ on } (-\infty,k) \text{ and } \eta \equiv 0 \text{ on } [k,\infty)\}$$

For each of the cases (i) and (ii), there is at most one extremal invariant measure with that property.

*Remark.*

(a) In the random environment case, (1.6) is satisfied if $[\delta(j)]^{-1}$ has finite moments of all orders. To see that some positivity assumption on the death rates is needed, consider the case in which the $\lambda$'s and $\rho$'s are all zero, and $\delta(j) = 0$ with positive probability. Then there would be uncountably many extremal invariant measures.

(b) It is possible for the number $N$ of distinct extremal invariant measures to have any of the values 1, 2, 3 or 4. To see this, consider the following examples.

$N = 1$ : A subcritical homogeneous contact process.

$N = 2$ : A supercritical homogeneous contact process.

$N = 3$ : Take $\delta(k) \equiv 1, \lambda(k) \equiv 4$, and $\rho(k) \equiv 1$ on the left half line and $\rho(k) \equiv 4$ on the right half line. In this case, there is an extremal invariant measure satisfying (i), but none satisfying (ii). To construct the measure satisfying (i), start the process with the configuration with $\eta(k) \equiv 1$ on the right half line and $\eta(k) \equiv 0$ on the left half line. Then on the left half line, the position of the leftmost one has a drift to the right, while on the right half line, the distribution of the process remains above the upper invariant measure for the one sided homogeneous contact process with parameter 4.

$N = 4$ : Take $\delta(k) \equiv 1, \lambda(k) \equiv 1$ on the right half line, $\lambda(k) \equiv 4$ on the left half line, $\rho(k) \equiv 1$ on the left half line, and $\rho(k) \equiv 4$ on the right half line. In this case, there is an invariant measure satisfying (i), and another one satisfying (ii).

## 2. Extinction.

In this section, we prove Theorems 1, 2, 3 and 4. Several techniques have been used to prove extinction in the case of the basic contact process. The proofs of the first three results are straightforward extensions of two of these.

*Proof of Theorem 1.* We may assume without loss that $\rho(j) > 0$ and $\lambda(j) > 0$ for all $j$. Let $\eta_t$ be the spatially inhomogeneous contact process, and let $\zeta_t$ be the modification of this process in which deaths at $k$ occur only if $\zeta(k-1) = 0$ or $\zeta(k+1) = 0$. These two processes can be coupled so that

$$\eta_t(k) \leq \zeta_t(k) \quad \text{for all } k. \tag{2.1}$$

If initially $\{k : \zeta(k) = 0\}$ is an interval, then $\{k : \zeta_t(k) = 0\}$ is an interval for all positive $t$. The boundaries of this interval evolve like independent birth and death chains until (possibly) the interval of zeros disappears when $\zeta$ is identically one. The rightmost zero makes transitions

$$n \to n+1 \text{ at rate } \delta(n+1) \text{ and } n \to n-1 \text{ at rate } \rho(n), \tag{2.2}$$

while the leftmost zero makes transitions

$$n \to n+1 \text{ at rate } \lambda(n) \text{ and } n \to n-1 \text{ at rate } \delta(n-1). \tag{2.3}$$

Therefore, if $m < 0 < n$ and we take $\zeta_0(k) = 0$ for $m \le k \le n$ and $\zeta_0(k) = 1$ for $k < m$ and $k > n$, it follows the known explicit formulas for hitting probabilities for birth and death chains that

$$\lim_{t\to\infty} P[\zeta_t(l) = 0] \ge \frac{\sum_{k=0}^{n-1} \prod_{j=0}^{k} \frac{\rho(j)}{\delta(j+1)}}{\sum_{k=0}^{\infty} \prod_{j=0}^{k} \frac{\rho(j)}{\delta(j+1)}} \, \frac{\sum_{k=m+1}^{0} \prod_{j=k}^{0} \frac{\lambda(j)}{\delta(j-1)}}{\sum_{k=-\infty}^{0} \prod_{j=k}^{0} \frac{\lambda(j)}{\delta(j-1)}}$$

for all $-\infty < l < \infty$. By assumption (1.3), the right hand side above tends to one as $m \to -\infty$ and $n \to +\infty$. Since $\delta(k) > 0$ for all $k$, for any $m < 0 < n$, there will be with probability one, times at which $\eta_t(k) = 0$ for all $m \le k \le n$. At such a time, one can begin the coupling with the process $\zeta_t$, preserving the inequality in (2.1). Therefore, for any initial distribution,

$$\lim_{t\to\infty} P[\eta_t(l) = 0] = 1$$

for all $l$, and hence $\eta_t$ dies out.

*Proof of Theorem 2.* Let $\mu$ be the upper invariant measure for the contact process, and set $\alpha(j) = \mu\{\eta : \eta(j) = 1\}$ and $\phi(k,l) = \mu\{\eta : \eta(k) = 1, \eta(l) = 1 \text{ and } \eta(j) = 0 \text{ for all } k < j < l\}$. Since $\mu$ is invariant, these functions satisfy the relations

$$\alpha(j) = \rho(j)\,[\alpha(j+1) - \phi(j,j+1)],$$

and

$$\begin{aligned} 2\phi(j,j+1) &= \rho(j)\,[\alpha(j+1) - \phi(j,j+1)] + \rho(j+1)\,\phi(j,j+2) \\ &\ge \rho(j)\,[\alpha(j+1) - \phi(j,j+1)]. \end{aligned}$$

Solving the second of these for $\phi(j, j+1)$ and using it in the first gives

$$\alpha(j) \leq \frac{2\rho(j)}{2+\rho(j)}\alpha(j+1).$$

Iterating this inequality and using the hypothesis of the theorem implies that $\alpha(j) = 0$ for each $j$.

*Proof of Theorem 3.* This proof begins much like the proof of Theorem 1. Couple together in the natural way the contact process in a random environment $\eta_t$ and Markov chains $R_t$ and $L_t$ with the transition rates given in (2.2) and (2.3) respectively so that $\eta_0 \equiv 1, R_0 = 0, L_0 = 1$, and $\eta_t(k) = 0$ whenever $L_t \leq k \leq R_t$. Since

$$P\{\eta_t(k) = 1\} \leq P\{L_t > k\} + P\{R_t < k\},$$

and the distribution of $P\{\eta_t(k) = 1\}$ is independent of $k$, it is enough to show that there is a deterministic $k_t$ so that the expected value over the environment of $P\{L_t > k_t\} + P\{R_t < k_t\}$ tends to zero as $t$ tends to infinity. Since $R_t$ and $L_t$ are random walks in a random environment, the continuous time version of Theorem 1.16 in Solomon (1975) implies that for almost every environment,

$$\lim_{t\to\infty}\frac{R_t}{t} = \frac{1 - E\left\{\rho(k)\left[\delta(k)\right]^{-1}\right\}}{E\left[\delta(k)\right]^{-1}},$$

and

$$\lim_{t\to\infty}\frac{L_t}{t} = \frac{1 - E\left\{\delta(k)\left[\lambda(k)\right]^{-1}\right\}}{E\left[\lambda(k)\right]^{-1}},$$

provided in each case that the expected values which appear on the right are finite and the numerator is strictly positive. Furthermore, if the numerator is less than or equal to zero, the limit is less than or equal to zero. Under the assumptions of Theorem 3, it is therefore possible to choose a $k_t$ as required.

*Proof of Theorem 4.* Let $\eta_t$ and $\zeta_t$ be as in the statement of the theorem, and let $\mu$ be the upper invariant measure for the process $\eta_t$. Then $\mu$ can be regarded as a function of the sequence $\{\rho(k), -\infty < k < \infty\}$. By attractiveness, it is an increasing function of this sequence, where the space of probability measures on $\{0,1\}^Z$ is given the usual partial order of stochastic monotonicity. The distribution of $\{\rho(k), -\infty < k < \infty\}$ has positive correlations, so it follows that

$$\begin{aligned}(2.4)\qquad & E\rho(k)\mu\{\eta : \eta(k+1) = 1 \text{ and } \eta(j) = 1 \text{ for some } j \in A\} \\ & \geq E\rho(k)\, E\mu\{\eta : \eta(k+1) = 1 \text{ and } \eta(j) = 1 \text{ for some } j \in A\}\end{aligned}$$

for all $k$ and any finite set $A$. (See Section 2 of Chapter II of Liggett (1985) for basic results on stochastic monotonicity and measures with positive correlations, and Harris (1960) for the original version on the result being used.) On the other hand, $\mu\{\eta : \eta(k+1) = 1\}$ is a function only of $\{\rho(j), j \geq k+1\}$, and is therefore independent of $\rho(k)$. Using this observation together with (2.4) gives

$$\begin{aligned} & E\rho(k)\mu\{\eta : \eta(k+1) = 1 \text{ and } \eta(j) = 0 \text{ for all } j \in A\} \\ &\quad = E\rho(k)\mu\{\eta : \eta(k+1) = 1\} \\ &\qquad - E\rho(k)\mu\{\eta : \eta(k+1) = 1 \text{ and } \eta(j) = 1 \text{ for some } j \in A\} \\ (2.5)\qquad &\quad \leq E\rho(k)E\mu\{\eta : \eta(k+1) = 1\} \\ &\qquad - E\rho(k)E\mu\{\eta : \eta(k+1) = 1 \text{ and } \eta(j) = 1 \text{ for some } j \in A\} \\ &\quad = E\rho(k)E\mu\{\eta : \eta(k+1) = 1 \text{ and } \eta(j) = 0 \text{ for all } j \in A\}.\end{aligned}$$

Since $\mu$ is invariant for the contact process in a random environment,

$$\begin{aligned} & \sum_{k\in A} \mu\{\eta : \eta(k) = 1, \eta \equiv 0 \text{ on } A\backslash\{k\}\} \\ (2.6)\qquad & \quad = \sum_{k\in A, k+1\notin A} \rho(k)\{\eta : \eta(k+1) = 1, \eta \equiv 0 \text{ on } A\}\end{aligned}$$

for all finite sets $A$. Define the probability measure $\nu$ on $\{0,1\}^Z$ by $\nu = E\mu$. Taking expected values in (2.6) and using (2.5) gives

$$\begin{aligned} & \sum_{k\in A} \nu\{\eta : \eta(k) = 1, \eta \equiv 0 \text{ on } A\backslash\{k\}\} \\ & \quad \leq \sum_{k\in A, k+1\notin A} E\rho(k)\nu\{\eta : \eta(k+1) = 1, \eta \equiv 0 \text{ on } A\}\end{aligned}$$

This can be re-expressed as

$$\frac{d}{dt}\nu_t\{\eta : \eta \equiv 0 \text{ on } A\} \leq 0 \tag{2.7}$$

at $t = 0$, where $\nu_t$ is the distribution at time $t$ of the process $\zeta_t$ with initial distribution $\nu$. Integrating the duality equation (1.2) with respect to $\nu_s$ gives

$$\nu_{t+s}\{\eta : \eta \equiv 0 \text{ on } A\} = \sum_B P^A[A_t = B]\,\nu_s\{\eta : \eta \equiv 0 \text{ on } B\}$$

where $A_t$ is the dual to $\zeta_t$. Therefore, (2.7) holds for positive $t$ as well. If $\zeta_t$ dies out, it follows that $\nu$ must be the pointmass on $\eta \equiv 0$, and hence that $\mu$ is the pointmass on $\eta \equiv 0$ almost surely. Therefore, $\eta_t$ dies out.

## 3. Survival.

We begin by defining an operator which will play an important role in our treatment of the survival problem for one dimensional spatially inhomogeneous contact processes. Given a positive function $\alpha$ on $Z$, define $\Gamma_\alpha\phi$ for functions $\phi(k,l)$, $-\infty < k < l < \infty$ by

$$\begin{aligned}\Gamma_\alpha\phi(k,k+1) = {} & \lambda(k)\,\phi(k-1,k+1) + \rho(k+1)\,\phi(k,k+2)\\ & + \lambda(k+1)\,\alpha(k) + \rho(k)\,\alpha(k+1)\\ & - [\delta(k) + \delta(k+1) + \rho(k) + \lambda(k+1)]\\ & \quad \times \phi(k,k+1),\end{aligned} \tag{3.1}$$

and

$$\begin{aligned}\Gamma_\alpha\phi(k,l) = {} & \lambda(k)\phi(k-1,l)\\ & + \rho(l)\phi(k,l+1)\\ & + \sum_{j=k+1}^{l-1}\frac{\delta(j)}{\alpha(j)}\phi(k,j)\phi(j,l)\\ & - [\delta(k) + \delta(l) + \lambda(k+1) + \rho(l-1)]\,\phi(k,l)\end{aligned} \tag{3.2}$$

for $k < l - 1$. To motivate this definition, suppose that $\alpha$ is strictly positive, $\phi$ is nonnegative, and $\alpha$ and $\phi$ satisfy

$$\sum_{k \le i < l} \phi(k, l) = 1, \tag{3.3}$$

and

$$\alpha(i) = \sum_{k > i} \phi(i, k) = \sum_{k > i} \phi(k, i) \tag{3.4}$$

for all $i \in Z$. We can then define the distribution $\mu$ of the inhomogeneous renewal process associated with $\alpha$ and $\phi$ as the probability measure on $\{0, 1\}^Z$ with cylinder probabilities given by

$$\begin{aligned} &\mu\{\eta : \eta(k_j) = 1 \text{ for } 1 \le j \le n, \eta(k) = 0 \text{ for all other } k_1 < k < k_n\} \\ &\qquad = \alpha(k_1) \prod_{i=1}^{n-1} \frac{\phi(k_i, k_{i+1})}{\alpha(k_i)} \end{aligned} \tag{3.5}$$

for $n \ge 1$ and $k_1 < k_2 < \cdots < k_n$. Identities (3.3) and (3.4) are needed for consistency, and in order to guarantee that $\mu$ concentrates on configurations with infinitely many ones in both directions. If $\mu_t$ is the distribution at time $t$ of the contact process with initial distribution $\mu$, then

$$\frac{d}{dt}\mu_t\{\eta : \eta(k) = \eta(l) = 1, \eta(j) = 0 \text{ for all } k < j < l\}\big|_{t=0} = \Gamma_\alpha \phi(k, l) \tag{3.6}$$

for all $k < l$. Thus the operator $\Gamma_\alpha$ would govern the evolution of the contact process if the distribution of the process were always a renewal measure. Identity (3.6) has the following important generalization.

**Proposition 3.7.** *Suppose $\alpha$ and $\phi$ satisfy (3.3) and (3.4), let $\mu$ be the measure determined by (3.5), and let $\mu_t$ be the distribution at time t of the contact process with initial distribution $\mu$. Fix a finite subset A of Z, and define*

$$R(k) = \mu\{\eta : \eta(j) = 0 \textit{ for all } j \in A \cap (k, \infty) \,|\, \eta(k) = 1\} \tag{3.8}$$

*and*

$$L(k) = \mu\{\eta : \eta(j) = 0 \textit{ for all } j \in A \cap (-\infty, k) \,|\, \eta(k) = 1\} \tag{3.9}$$

*for* $k \in Z$. *Then*

$$\begin{aligned}
&\frac{d}{dt}\mu_t\{\eta : \eta(j) = 0 \textit{ for all } j \in A\}\big|_{t=0} \\
&\quad = \sum_{k<1;k,l\notin A} L(k)R(l)\Gamma_\alpha\pi(k,l) \\
&\qquad - \sum_{j\notin A} L(j)R(j)\frac{d}{dt}\mu_t\{\eta : \eta(j) = 1\}\big|_{t=0} \\
&\qquad - \sum_{k\in A,k+1\notin A} \lambda(k+1)\alpha(k)L(k+1)[R(k) - R(k+1)] \\
&\qquad - \sum_{k\notin A,k+1\in A} \rho(k)\alpha(k+1)R(k)[L(k+1) - L(k)] \\
&\qquad - \sum_{k,k+1\notin A} [\lambda(k+1)\alpha(k) + \rho(k)\alpha(k+1)][L(k+1) - L(k)] \\
&\qquad\qquad \cdot [R(k) - R(k+1)]
\end{aligned} \tag{3.10}$$

*where in the first sum, k is restricted to be to the left of the rightmost point in A and l is restricted to be to the right of the leftmost point in A, while in the second sum, j is restricted to be between the leftmost and rightmost points in A. (Note that only finitely many nonzero terms appear in the other three sums on the right hand side of (3.10), since* $L(k) = 1$ *if k is to the left of A, and* $R(k) = 1$ *if k is to the right of A.)*

*Proof.* Using the dynamics of the contact process and the renewal property of $\mu$, we see that

$$\begin{aligned}
&\frac{d}{dt}\mu_t\{\eta : \eta(j) = 0 \text{ for all } j \in A\}\big|_{t=0} \\
&\quad = \sum_{j\in A} \delta(j)\,\alpha(j)\,L(j)\,R(j)
\end{aligned}$$

$$- \sum_{k\notin A,k+1\in A} \lambda(k+1)\,\alpha(k)\,L(k)\,R(k)$$
$$- \sum_{k\in A,k+1\notin A} \rho(k)\,\alpha(k+1)\,L(k+1)\,R(k+1)\,.$$

The renewal property of $\mu$ also gives

$$\alpha(j)\,L(j) = \sum_{k<j,k\notin A} L(k)\,\phi(k,j) \tag{3.11}$$

and

$$\alpha(j)\,R(j) = \sum_{l>j,l\notin A} R(l)\,\phi(j,l)\,. \tag{3.12}$$

Therefore, letting $a = \min\{j : j \in A\}$ and $b = \max\{j : j \in A\}$, we see that

$$\begin{aligned}
&\sum_{j\in A} \delta(j)\alpha(j)L(j)R(j) \\
&\quad= \sum_{\substack{k<j<l \\ j\in A;k,l\notin A}} L(k)\phi(k,j)\phi(j,l)R(l)\frac{\delta(j)}{\alpha(j)} \\
&\quad= \sum_{\substack{k<l;k,l\notin A \\ k<b;l>a}} L(k)R(l) \sum_{k<j<l} \frac{\delta(j)}{\alpha(j)}\phi(k,j)\phi(j,l) \\
&\qquad - \sum_{\substack{k<l;k,l\notin A \\ k<b;l>a}} L(k)R(l) \sum_{k<j<l,j\notin A} \frac{\delta(j)}{\alpha(j)}\phi(k,j)\phi(j,l) \\
&\quad= \sum_{\substack{k<l;k,l\notin A \\ k<b;l>a}} L(k)R(l)\{\Gamma_\alpha\phi(k,l) - \lambda(k)\phi(k-1,l) - \rho(l)\phi(k,l+1) \\
&\qquad + [\delta(k) + \delta(l) + \lambda(k+1) + \rho(l-1)]\,\phi(k,l)\} \\
&\qquad - \sum_{\substack{k,k+1\notin A \\ a<k<b}} L(k)R(k+1)\,[\lambda(k+1)\alpha(k) + \rho(k)\alpha(k+1)] \\
&\qquad - \sum_{\substack{k<l;k,l\notin A \\ k<b;l>a}} L(k)R(l) \sum_{k<j<l,j\notin A} \frac{\delta(j)}{\alpha(j)}\phi(k,j)\phi(j,l)
\end{aligned} \tag{3.13}$$

To get the final equality, we have used the definition of $\Gamma_\alpha$. The middle sum on the right hand side of (3.13) comes from the fact that the definition of $\Gamma_\alpha \phi(k,l)$ is slightly different in the case $l = k+1$ from the case $l > k+1$. Using (3.11) and (3.12) again, note that the terms on the right of (3.13) which involve $\delta$'s combine to give

$$\sum_{a<j<b, j\notin A} \delta(j)\,\alpha(j)\,L(j)\,R(j)$$

On the other hand,

$$\frac{d}{dt}\mu_t\{\eta : \eta(j) = 1\}\big|_{t=0} = \lambda(j)\,[\alpha(j-1) - \phi(j-1,j)] \\ + \rho(j)\,[\alpha(j+1) - \phi(j,j+1)] - \delta(j)\,\alpha(j).$$

Therefore, the difference between the left side of (3.10) and the first two terms on the right side is equal to

$$\begin{aligned} &\sum_{\substack{k<l;k,l\notin A\\ k<b;l>a}} L(k)R(l)\{[\lambda(k+1)+\rho(l-1)]\phi(k,l) \\ &\qquad -\lambda(k)\phi(k-1,l) - \rho(l)\phi(k,l+1)\} \\ &+ \sum_{a<j<b,j\notin A} L(j)R(j)\{\lambda(j)[\alpha(j-1)-\phi(j-1,j)] \\ &\qquad +\rho(j)[\alpha(j+1)-\phi(j,j+1)]\} \\ &- \sum_{k,k+1\notin A,a<k<b} L(k)R(k+1)[\lambda(k+1)\alpha(k)+\rho(k)\alpha(k+1)] \\ &- \sum_{k\notin A,k+1\in A} \lambda(k+1)\alpha(k)L(k)R(k) \\ &- \sum_{k\in A,k+1\notin A} \rho(k)\alpha(k+1)L(k+1)R(k+1). \end{aligned} \tag{3.14}$$

Making appropriate changes of variables in the sums and using identities (3.11) and (3.12),we see that the terms in (3.14) which involve $\phi$'s combine to give

$$\sum_{\substack{k<l;k,l\notin A\\ k<b;l>a}} L(k)R(l)\lambda(k+1)\phi(k,l)$$

$$
\begin{aligned}
&- \sum_{\substack{k<l;k+1,l\notin A\\ k+1<b;l>a}} L(k+1)R(l)\lambda(k+1)\phi(k,l)\\
&+ \sum_{\substack{k\le l;k,l+1\notin A\\ k<b;l+1>a}} L(k)R(l+1)\rho(l)\phi(k,l+1)\\
&- \sum_{\substack{k\le l;k,l\notin A\\ k<b;l>a}} L(k)R(l)\rho(l)\phi(k,l+1)\\
= &\sum_{a\le k+1\le b,k\notin A} \lambda(k+1)\alpha(k)R(k)L(k)\\
&- \sum_{a\le k+1\le b,k+1\notin A} \lambda(k+1)\alpha(k)R(k)L(k+1)\\
&+ \sum_{a\le l\le b,l+1\notin A} \rho(l)\alpha(l+1)L(l+1)R(l+1)\\
&- \sum_{a\le l\le b,l\notin A} \rho(l)\alpha(l+1)L(l+1)R(l).
\end{aligned}
$$

Using this in (3.14), it follows that the difference between the left side of (3.10) and the first two terms on the right side is equal to

$$
\begin{aligned}
&\sum_{a<k+1<b;k,k+1\notin A} \lambda(k+1)\alpha(k)L(k)[R(k)-R(k+1)]\\
&+ \sum_{a<k+1<b,k+1\notin A} \lambda(k+1)\alpha(k)L(k+1)[R(k+1)-R(k)]\\
&+ \sum_{a<k<b;k,k+1\notin A} \rho(k)\alpha(k+1)R(k+1)[L(k+1)-L(k)]\\
&- \sum_{a<k<b,k\notin A} \rho(k)\alpha(k+1)R(k)[L(k)-L(k+1)].
\end{aligned}
$$

This is what is required to complete the proof, since $L(k)=1$ if $k\le a$ and $R(k)=1$ if $k\ge b$.

**Theorem 3.15.** *Suppose that there exist functions $\alpha > 0$ and $\phi \geq 0$ which satisfy (3.3), (3.4), $\Gamma_\alpha \phi \equiv 0$, and*

$$\phi(k,l)\,\phi(k-1,l+1) \geq \phi(k,l+1)\,\phi(k-1,l) \tag{3.16}$$

*for all $k < 1$. Then the infinite contact process survives.*

*Proof.* By the FKG theorem (see Corollary 2.12 of Chapter II in Liggett (1985)) and (3.16), the renewal measure $\mu$ given in (3.5) has positive correlations. The same is true for the conditioned renewal measure $\mu(\cdot\ |\eta(k)=1)$. Take a finite set $A$ and define $R(k)$ and $L(k)$ as in Proposition 3.7. By the renewal property,

$$R(k+1) = \mu\{\eta : \eta(j) = 0 \text{ for all } j \in A \cap (k,\infty) \mid \eta(k) = 1, \eta(k+1) = 1\}$$

if $k+1 \notin A$, so it follows from the positive correlations property that $R(k+1) \leq R(k)$ in this case. Similarly, $L(k) \leq L(k+1)$ if $k \notin A$. Therefore, the last three sums on the right side of (3.10) are nonnegative. Since $\Gamma_\alpha \phi \equiv 0$, the first sum is zero. By summing (3.6) for $l > k$, we see that the second term on the right side of (3.10) is zero as well. Thus by Proposition 3.7,

$$\frac{d}{dt}\mu_t\{\eta : \eta(j) = 0 \text{ for all } j \in A\}\,|_{t=0} \leq 0.$$

It then follows from the duality relation (1.2) as in the proof of Theorem 4 that

$$\frac{d}{dt}\mu_t\{\eta : \eta(j) = 0 \text{ for all } j \in A\} \leq 0$$

for all $t \geq 0$ and all finite sets $A$. Therefore, the upper invariant measure $\nu$ satisfies

$$\nu\{\eta : \eta(j) = 0 \text{ for all } j \in A\} \leq \mu\{\eta : \eta(j) = 0 \text{ for all } j \in A\},$$

and hence the contact process survives.

The next proposition is useful in checking condition (3.16) in special cases. In order to avoid technical difficulties, we will assume in this result that the $\lambda$'s, $\rho$'s, and $\delta$'s are uniformly bounded.

**Proposition 3.17.** *Suppose that there exist functions $\alpha > 0$ and $\phi \geq 0$ which satisfy (3.3), (3.4) and $\Gamma_\alpha \phi \equiv 0$. Define continuous time Markov chains $X_t$ and $Y_t$ on $Z$ as follows: $X_t$ goes from $k$ to $k+1$ at rate $\lambda(k+1)$, and from $k$ to $m < k$ at rate $\delta(k)\phi(m,k)/\alpha(k)$. $Y_t$ goes from $k$ to $k-1$ at rate $\rho(k-1)$, and from $k$ to $m > k$ at rate $\delta(k)\phi(k,m)/\alpha(k)$. Suppose that*

$$\sum_{m \geq k} P^m(X_t = k \text{ for some } t \geq 0) < \infty$$

*and*

$$\sum_{m \leq k} P^m(Y_t = k \text{ for some } t \geq 0) < \infty$$

*for each $k$. Then (3.16) is satisfied.*

*Proof.* Let $B$ be the Banach space

$$B = \left\{ \psi : \|\psi\| = \sup_{k<l} \frac{|\psi(k,l)|}{\phi(k,l)} < \infty \right\}$$

and $B_1$ its unit ball. Since the transition rates have been assumed to be bounded, $\Gamma_\alpha$ is Lipschitz continuous on $B_1$ . In order to see that, write

$$\begin{aligned}
&(\Gamma_\alpha \psi_1 - \Gamma_\alpha \psi_2)(k,l) \\
&\quad = \lambda(k)\left[\psi_1(k-1,l) - \psi_2(k-1,l)\right] \\
&\qquad + \rho(l)\left[\psi_1(k,l+1) - \psi_2(k,l+1)\right] \\
&\qquad + \sum_{j=k+1}^{l-1} \frac{\delta(j)}{\alpha(j)} \{\psi_1(k,j)\left[\psi_1(j,l) - \psi_2(j,l)\right] \\
&\qquad\quad + \psi_2(j,l)\left[\psi_1(k,j) - \psi_2(k,j)\right]\} \\
&\qquad - \left[\delta(k) + \delta(l) + \rho(l-1) + \lambda(k+1)\right]\left[\psi_1(k,l) - \psi_2(k,l)\right]
\end{aligned}$$

so that the Lipschitz norm of $\Gamma_\alpha$ on $B_1$ is bounded by

$$\begin{aligned}
\sup_{k<l} \Bigg\{ & \frac{\lambda(k)\,\phi(k-1,l)}{\phi(k,l)} + \frac{\rho(l)\,\phi(k,l+1)}{\phi(k,l)} \\
& + 2\sum_{j=k+1}^{l-1} \frac{\delta(j)\,\phi(k,j)\,\phi(j,l)}{\alpha(j)\,\phi(k,l)} + \left[\delta(k) + \delta(l) + \lambda(k+1) + \rho(l-1)\right] \Bigg\}
\end{aligned}$$

which is finite since $\phi$ satisfies $\Gamma_\alpha \phi \equiv 0$, and the rates are bounded. A similar argument shows that $\Gamma_\alpha$ is dissipative, and then that it generates a semigroup $S(t)$ of nonlinear contractions on $B_1$. (See Pavel (1987), for example, for a treatment of nonlinear semigroup theory.) $\Gamma_\alpha$ plus a large multiple of the identity is a positive increasing operator, so $S(t)$ has these properties as well. Using this semigroup, we may define $\psi_t(k,l)$ for $k < l$ by $\psi_0(k,l) \equiv 0$ and

$$\frac{d}{dt}\psi_t = \Gamma_\alpha \psi_t. \tag{3.18}$$

Then $\psi_t(k,l)$ is nonnegative and nondecreasing in $t$ for each $k,l$. Therefore, we may let

$$\psi(k,l) = \lim_{t\to\infty} \psi_t(k,l) \le \phi(k,l)$$

Passing to the limit in (3.18), we see that $\Gamma_\alpha \psi \equiv 0$ as well. So , $\Delta(k,l) = \phi(k,l) - \psi(k,l)$ satisfies

$$\begin{aligned} 0 = {} & \lambda(k)\,\Delta(k-1,l) + \rho(l)\,\Delta(k,l+1) \\ & + \sum_{j=k+1}^{l-1} \frac{\delta(j)}{\alpha(j)} \left[\phi(k,j)\,\Delta(j,l) + \Delta(k,j)\,\phi(j,l) - \Delta(k,j)\,\Delta(j,l)\right] \\ & - \left[\delta(k) + \delta(l) + \lambda(k+1) + \rho(l-1)\right]\Delta(k,l). \end{aligned}$$

Comparing this with the generator of the Markov chain $(X_t, Y_t)$, where $X_t$ and $Y_t$ are independent copies of the chains defined in the statement of the theorem, we see that

$$\frac{d}{dt}\sum_{k<l} \Delta(k,l)\,P^k(X_t = m)\,P^l(Y_t = n) \ge 0$$

for all $m$ and $n$ at $t = 0$, and hence at all $t \ge 0$. Therefore,

$$\sum_{k<l} \Delta(k,l)\,P^k(X_t = m)\,P^l(Y_t = n) \ge \Delta(m,n)$$

for $m < n$. From this we concluded that $\Delta \equiv 0$ by letting $t \to \infty$, using the hypothesis of then theorem and dominated convergence. The domination comes from

$$\begin{aligned}
&\sum_{k<l} \Delta(k,l) P^k(X_t = m) P^l(Y_t = n) \\
&\qquad \leq \sum_{k<m, l>n} \phi(k,l) + \sum_{k \geq m} P^k(X_s = m \text{ for some } s \geq 0)\alpha(k) \\
&\qquad\qquad + \sum_{l \leq n} P^l(Y_s = n \text{ for some } s \geq 0)\alpha(l),
\end{aligned}$$

which is finite by the assumption and (3.3). So, $\psi \equiv \phi$ , and hence it suffices to prove that $\psi_t$ satisfies (3.16) for each $t \geq 0$. To do so, compute

$$\begin{aligned}
&\frac{d}{dt}[\psi_t(n-1,n+2)\psi_t(n,n+1) - \psi_t(n-1,n+1)\psi_t(n,n+2)] \\
&= \lambda(n-1)[\psi_t(n-2,n+2)\psi_t(n,n+1) - \psi_t(n,n+2)\psi_t(n-2,n+1)] \\
&\qquad + \lambda(n+1)\alpha(n)\psi_t(n-1,n+2) \\
&\qquad + \rho(n+2)[\psi_t(n-1,n+3)\psi_t(n,n+1) - \psi_t(n,n+3)\psi_t(n-1,n+1)] \\
&\qquad + \rho(n)\alpha(n+1)\psi_t(n-1,n+2) \\
&\qquad - [\psi_t(n-1,n+2)\psi_t(n,n+1) - \psi_t(n-1,n+1)\psi_t(n,n+2)] \\
&\qquad\qquad \cdot [\lambda(n) + \lambda(n+1) + \rho(n) + \rho(n+1) \\
&\qquad\qquad\qquad + \delta(n-1) + \delta(n) + \delta(n+1) + \delta(n+2)]
\end{aligned}$$

and

$$\begin{aligned}
&\frac{d}{dt}[\psi_t(k-1,l+1)\psi_t(k,l) - \psi_t(k-1,l)\psi_t(k,l+1)] \\
&= \lambda(k-1)[\psi_t(k-2,l+1)\psi_t(k,l) - \psi_t(k,l+1)\psi_t(k-2,l)] \\
&\qquad + \rho(l+1)[\psi_t(k-1,l+2)\psi_t(k,l) - \psi_t(k,l+2)\psi_t(k-1,l)] \\
&\qquad + \frac{\psi_t(k-1,l+1)\psi_t(k,l) - \psi_t(k-1,l)\psi_t(k,l+1)}{\psi_t(k,l)} \\
&\qquad\quad \cdot \sum_{j=k+1}^{l-1} \frac{\delta(j)}{\alpha(j)} \psi_t(k,j)\psi_t(j,l)
\end{aligned}$$

$$+\sum_{j=k+1}^{l-1}\frac{\delta(j)}{\alpha(j)\psi_t(k,l)}[\psi_t(k-1,l)\psi_t(k,j)-\psi_t(k,l)\psi_t(k-1,j)]$$
$$\cdot[\psi_t(k,l+1)\psi_t(j,l)-\psi_t(k,l)\psi_t(j,l+1)]$$
$$-[\psi_t(k-1,l+1)\psi_t(k,l)-\psi_t(k-1,l)\psi_t(k,l+1)]$$
$$\times[\lambda(k)+\lambda(k+1)+\rho(l)+\rho(l-1)$$
$$+\delta(k-1)+\delta(k)+\delta(l)+\delta(l+1)]$$

for $l > k+1$. Note that if $\psi_t$ satisfies (3.16) for a given $t$, then all terms in each of the above expressions except the last is nonnegative. This makes it clear that the semigroup $S(t)$ maps the class $P$ of nonnegative functions in $B_1$ which satisfy (3.16) into itself. Since the zero function is in $P$, that will conclude the proof. To carry out the details of the proof of the invariance of $P$ under $S(t)$, it is simplest to prove first that $P$ is invariant under the semigroups $S_1(t)$ and $S_2(t)$ whose generators are given respectively by the terms of $\Gamma_\alpha$ which contain only $\delta$'s , and those which contain only $\lambda$'s and $\rho$'s. The invariance of $P$ under $S(t)$ then follows from a nonlinear version of Trotter's product formula:

$$S(t)\psi=\lim_{n\to\infty}[S_1(t/n)S_2(t/n)]^n\psi.$$

(See Corollary 4.3 of Brezis and Pazy (1972), for example.) To show the invariance of $P$ under $S_1(t)$, write $\psi_t = S_1(t)\psi$ for some $\psi$ in $P$, and note that

$$\frac{d}{dt}[\psi_t(k-1,l+1)\psi_t(k,l)-\psi_t(k-1,l)\psi_t(k,l+1)]$$
$$=\frac{\psi_t(k-1,l+1)\psi_t(k,l)-\psi_t(k-1,l)\psi_t(k,l+1)}{\psi_t(k,l)}$$
$$\cdot\sum_{j=k+1}^{l-1}\frac{\delta(j)}{\alpha(j)}\psi_t(k,j)\psi_t(j,l)$$
$$+\sum_{j=k+1}^{l-1}\frac{\delta(j)}{\alpha(j)\psi_t(k,l)}[\psi_t(k-1,l)\psi_t(k,j)-\psi_t(k,l)\psi_t(k-1,j)]$$
$$\cdot[\psi_t(k,l+1)\psi_t(j,l)-\psi_t(k,l)\psi_t(j,l+1)]$$
$$-[\psi_t(k-1,l+1)\psi_t(k,l)-\psi_t(k-1,l)\psi_t(k,l+1)]$$
$$\cdot[\delta(k-1)+\delta(k)+\delta(l)+\delta(l+1)]$$

for $l > k$. Therefore, one can argue inductively in $l - k$ that (1.16) holds for $t > 0$. A similar argument works for $S_2(t)$ if only finitely many of the $\lambda$'s and $\rho$'s are nonzero, and one can then pass to a limit to handle the general case.

Next, we consider two examples in which the conditions of Theorem 3.15 can be checked by finding the required $\phi$ explicity. They are followed by other examples which give numerical results in cases in which explicit computation is not feasible. The first example is the translation invariant one from Holley and Liggett (1978). See also Section VI.1 of Liggett (1985).

*Example 3.19.* Take $\lambda(k) \equiv \lambda, \rho(k) \equiv \rho$, and $\delta(k) \equiv \delta$, where $\lambda + \rho \geq 4\delta$. Then

$$\alpha(k) \equiv \frac{1}{2} + \sqrt{\frac{1}{4} - \frac{\delta}{\lambda + \rho}}$$

and

$$\phi(k, k+n) = \frac{(2n-2)!}{(n-1)!(n+1)!}\left(\frac{\delta}{\lambda+\rho}\right)^n \cdot \left\{\left(\frac{\lambda+\rho}{\delta} - 4\right)n + \frac{\lambda+\rho}{\delta} + 2\right\}\left\{\frac{1}{2} + \sqrt{\frac{1}{4} - \frac{\delta}{\lambda+\rho}}\right\}.$$

This expression is sufficiently simple that one can check (3.16) directly.

In the next example, we consider the case in which the process is periodic of period 2, and is also invariant under the transformation $\{\eta(k), k \in Z\} \to \{\eta(-k+1), k \in Z\}$, which makes it self-dual. Note that nothing is lost in the result in this case by comparison with the translation invatiant case in Example 3.19.

*Example 3.20.* Take $\lambda(2n+1) = \lambda_1$, $\lambda(2n) = \lambda_2$, $\rho(2n) = \lambda_1$, $\rho(2n+1) = \lambda_2$, and $\delta(n) = \delta$. Then the infinite contact process survives provided that $\lambda_1\lambda_2 \geq 4\delta^2$.

*Proof.* Without loss of generality, we may take $\delta = 1$. In order to apply Theorem 3.15, we will look for a $\phi$ which is invariant under the same transformations as is the process itself: $\alpha(n) \equiv \alpha, \phi(k, k+2n)$ is independent of $k$, and $\phi(k, k+2n+1)$ is periodic in $k$ of period 2. Define the following generating functions:

$$F(s) = \sum_{n=1}^{\infty} \phi(k, k+2n)s^n, \text{ and } G_k(s) = \sum_{n=1}^{\infty} \phi(k, k+2n+1)s^n.$$

The equations $\Gamma_\alpha \phi \equiv 0$ then become

$$\begin{aligned} 0 &= F^2(s) + sG_1(s)G_2(s) + \lambda_1\alpha G_2(s) \\ &\quad + \lambda_2\alpha G_1(s) - (2+\lambda_1+\lambda_2)\alpha F(s) - (\lambda_1+\lambda_2-1)\alpha^2, \\ 0 &= sF(s)G_1(s) + \lambda_1\alpha F(s) + \lambda_2\alpha^2 s - (1+\lambda_2)\alpha s G_1(s) \end{aligned} \tag{3.21}$$

and

$$0 = sF(s)G_2(s) + \lambda_2\alpha F(s) + \lambda_1\alpha^2 s - (1+\lambda_1)\alpha s G_2(s). \tag{3.22}$$

Solving the last two equations for $G_1(s)$ and $G_2(s)$ respectively, and then substituting into the first equation yields the following equation for $F(s)$:

$$\begin{aligned} 0 = &F^4(s) - 2\alpha(2+\lambda_1+\lambda_2)F^3(s) \\ &+ \alpha^2\left\{(2+\lambda_1+\lambda_2)^2 + 2 + \lambda_1\lambda_2 - \lambda_1\lambda_2 s^{-1}\right\}F^2(s) \\ &- \alpha^3(2+\lambda_1+\lambda_2)\left\{2+\lambda_1\lambda_2 - \lambda_1\lambda_2 s^{-1}\right\}F(s) \\ &+ \alpha^4\left\{\lambda_1\lambda_2 s + 1 - \lambda_1\lambda_2\right\}. \end{aligned}$$

This can be rewritten as

$$\begin{aligned} 0 = &\left[F^2(s) - \alpha(2+\lambda_1+\lambda_2)F(s) + \alpha^2\right]^2 \\ &+ \alpha^2\lambda_1\lambda_2(1-s^{-1})\left[F^2(s) - \alpha(2+\lambda_1+\lambda_2)F(s) + \alpha^2\right] \\ &+ \alpha^4\lambda_1\lambda_2(s-1)^2 s^{-1}. \end{aligned}$$

Solving quadratic equations twice, and using the choice of sign which makes $F(s)$ well defined on [0,1] and satisfy $F(0) = 0$ leads to

$$\begin{aligned} 2\alpha^{-1}F(s) = &\, 2 + \lambda_1 + \lambda_2 \\ &- \sqrt{(4+\lambda_1+\lambda_2)(\lambda_1+\lambda_2) + \frac{2\lambda_1\lambda_2(1-s)}{s}\left(1 - \sqrt{1 - \tfrac{4s}{\lambda_1\lambda_2}}\right)}. \end{aligned}$$

Using (3.21) and (3.22), we see that we have written $F(s), G_1(s)$ and $G_2(s)$, and hence $\phi(k,l)$, in terms of $\alpha$ , provided that the expression above is a power series in $s$ whose radius of convergence is at least one. This is the case if and only if $\lambda_1\lambda_2 \geq 4$, which we will assume from now on. The value of $\alpha$ is determined by (3.3). Next we must check that $\phi(k,l) \geq 0$ for all $k$ and $l$. To do so, use the power series

$$1-\sqrt{1-x}=2\sum_{n=1}^{\infty}\frac{(2n-2)!}{n!(n-1)!}\left(\frac{x}{4}\right)^n$$

to rewrite

$$\alpha^{-1}F(s)=(2+\lambda_1+\lambda_2)\sum_{m=1}^{\infty}\frac{(2m-2)!}{m!(m-1)!}\cdot\left\{\sum_{n=1}^{\infty}\frac{(2n-2)!\,[(n+1)\lambda_1\lambda_2+2-4n]\,s^n}{(n-1)!(n+1)!(2+\lambda_1+\lambda_2)^2(\lambda_1\lambda_2)^n}\right\}^m$$

which is a power series with positive coefficients provided that $\lambda_1\lambda_2 \geq 4$. From (3.21) and (3.22) it then follows that $G_1(s)$ and $G_2(s)$ have power series with positive coefficients. Therefore, $\phi(k,l) \geq 0$ for all $k$ and 1. Finally, we must check (3.16) in order to conclude that the contact process survives. By continuity, it is enough to check this when $\lambda_1\lambda_2 > 4$, which will be assumed from now on. We will apply Proposition 3.17. Using the same notation as in that result, note first that $Y_t$ has the same transition law as $-X_t + 1$. Therefore it is enough to verify the assumption for $X_t$. Suppose $X_0$ is even, and let $0 = \tau_0 < \tau_1 < \cdots < \tau_n < \cdots$ be the succesive times at which $X_t$ is even. Then

$$X_{\tau_n}$$

is a random walk. The assumption of Proposition 3.17 will be satisfied provided that the increments of this random walk have a positive mean and finite second moment.

The second moment is automatically finite when $\lambda_1\lambda_2 > 4$, since then the power series for $F(s), G_1(s)$ and $G_2(s)$ have radius of convergence greater than one, and hence $\phi(k, k+n)$ decays exponentially rapidly in $n$. The mean of the increment is

$$\frac{\alpha\lambda_1 - 2F'(1) - 2G_1'(1) - G_1(1)}{\alpha(1+\lambda_1)} \tag{3.23}$$
$$+\frac{[\alpha\lambda_2 - 2F'(1) - 2G_2'(1) - G_2(1)]\,[\alpha(1+\lambda_1) - F(1)]}{\alpha(1+\lambda_1)\,[\alpha(1+\lambda_2) - F(1)]}.$$

Using (3.21) and (3.22), compute the first derivative of $G_1$ and $G_2$ in terms of $F$ and its derivative. Then use the expression obtained above for $F(s)$ to compute $F(1)$ and $F'(1)$, and substitute into (3.23) to get

$$\frac{4\lambda_1\lambda_2\sqrt{1-\frac{4}{\lambda_1\lambda_2}}}{(1+\lambda_1)\left(\lambda_2-\lambda_1+\sqrt{(4+\lambda_1+\lambda_2)(\lambda_1+\lambda_2)}\right)},$$

which is strictly positive as required.

*Example 3.24.* Consider the case in which $\lambda(k) \equiv 2, \rho(k) \equiv 2, \delta(2k) \equiv \delta(0)$, and $\delta(2k+1) \equiv \delta(1)$. Computer calculations indicate that for each of the choices of $\delta(0)$ in the first column below, the largest value of $\delta(1)$ for which the assumptions of Theorem 5 are satisfied is approximately that given in the second column:

| $\delta(0)$ | $\delta(1)$ |
|---|---|
| 1 | 1.000 |
| 2 | 0.490 |
| 4 | 0.226 |
| 8 | 0.095 |
| 16 | 0.035 |
| 32 | 0.011 |

As suggested by the above numbers, explicit calculations show that there is no positive constant $\kappa$ with the property that the assumptions of Theorem 5 are satisfied whenever $\delta(0)\delta(1) \leq \kappa$. Nevertheless, using the method described in his paper in this volume, Durrett has shown that there is a (very small) positive $\kappa$ with the property that the process survives whenever $\delta(0)\delta(1) \leq \kappa$.

A comparison of the results in Examples 3.20 and 3.24 suggests that it is more likely that one can find natural sufficient conditions on the rates for

Theorem 5 to be applicable when the birth rates are not constant than when the death rates are not constant. We conclude this section by reporting on some computer calculations which are intended to suggest when one might expect the conditions of Theorem 5 to be satisfied if the death rates are constant but the birth rates are chosen randomly.

*Example 3.25.* Take $\delta(k) = \delta$, and $0 \leq \lambda(k+1) = \rho(k) \leq 1$ for all $k$. Define sequences $\alpha_n(k)$ and $\phi_n(k,l)$ by $\alpha_0(k) \equiv 1$, $\phi_0(k, k+1) \equiv 1$, $\phi_0(k,l) = 0$ for $l > k+1$,

$$2(1+\delta)\alpha_{n+1}(k) = (2+\delta)\alpha_n(k) + \rho(k)[\alpha_n(k+1) - \phi_n(k,k+1)] + \lambda(k)[\alpha_n(k-1) - \phi_n(k-1,k)],$$

$$\begin{aligned}2(1+\delta)\phi_{n+1}(k,k+1) = {} & \rho(k+1)\phi_n(k,k+2) + \lambda(k)\phi_n(k-1,k+1) \\ & + \rho(k)\alpha_n(k+1) + \lambda(k+1)\alpha_n(k) \\ & + [2 - \rho(k) - \lambda(k+1)]\phi_n(k,k+1),\end{aligned}$$

and

$$\begin{aligned}2(1+\delta)\phi_{n+1}(k,l) = {} & \rho(l)\phi_n(k,l+1) + \lambda(k)\phi_n(k-1,l) \\ & + [2 - \rho(l-1) - \lambda(k+1)]\phi_n(k,l) \\ & + \delta \sum_{j=k+1}^{l-1} \frac{\phi_n(k,j)\phi_n(j,l)}{\alpha_n(j)}\end{aligned}$$

for $l > k+1$. Then $\alpha_n(k)$ and $\phi_n(k,l)$ satisfy (3.3) and (3.4) for all $n$, and the limits as $n \to \infty$, if they exist, will satisfy $\Gamma_\alpha \phi \equiv 0$. Thus, one would expect the conditions of Theorem 5 to be satisfied whenever these limits are not identically zero. Based on the result in Example 3.19, one would expect the limiting $\alpha$, as a function of $\delta$, to have (negative) infinite derivatives of all orders at the largest value of $\delta$ at which $\alpha$ is not zero. So, it is reasonable to take as an estimate of $\delta^*$, the largest $\delta$ for which the conditions of Theorem 5 are satisfied, that $\delta$ which maximizes the negative of the first or second derivative of $\alpha_n(k)$ with respect to $\delta$, for a large value of $n$, say $n = 75$. If $\lambda(k) \equiv .5$, we know from Example 3.19 that $\delta^* = .25$. The estimates for $\delta^*$ obtained using the above procedure are .30 based on the first derivative and .22 based on the second derivative. Thus in this case, the

| | | | | |
|---|---|---|---|---|
| (.25,.30) | (.26,.30) | (.26,.30) | (.25,.30) | (.23,.30) |
| (.30,.30) | (.28,.30) | (.22,.30) | (.20,.30) | (.19,.30) |
| (.25,.30) | (.25,.30) | (.25,.30) | (.24,.30) | (.22,.30) |
| (.23,.30) | (.25,.30) | (.25,.30) | (.26,.30) | (.26,.30) |
| (.22,.30) | (.22,.30) | (.22,.30) | (.21,.30) | (.21,.30) |
| (.18,.27) | (.22,.29) | (.24,.30) | (.26,.30) | (.26,.30) |
| (.20,.29) | (.19,.28) | (.23,.30) | (.23,.30) | (.24,.30) |
| (.23,.30) | (.23,.30) | (.22,.30) | (.21,.30) | (.17,.27) |
| (.22,.30) | (.21,.30) | (.20,.29) | (.19,.27) | (.18,.27) |
| (.20,.29) | (.20,.27) | (.19,.27) | (.20,.28) | (.21,.30) |
| (.17,.24) | (.18,.26) | (.20,.30) | (.20,.30) | (.20,.30) |
| (.21,.29) | (.20,.29) | (.19,.28) | (.17,.26) | (.17,.26) |
| (.19,.28) | (.19,.26) | (.18,.25) | (.18,.27) | (.19,.29) |
| (.13,.22) | (.18,.26) | (.19,.27) | (.19,.29) | (.19,.29) |
| (.16,.25) | (.16,.27) | (.17,.27) | (.17,.26) | (.18,.26) |
| (.14,.20) | (.15,.23) | (.16,.25) | (.16,.25) | (.17,.25) |
| (.19,.28) | (.12,.23) | (.10,.22) | (.10,.21) | (.16,.25) |
| (.16,.25) | (.16,.25) | (.13,.23) | (.11,.22) | (.13,.17) |
| (.16,.24) | (.14,.20) | (.10,.18) | (.10,.20) | (.10,.19) |
| (.15,.22) | (.10,.18) | (.10,.18) | (.10,.18) | (.10,.17) |
| (.15,.23) | (.14,.21) | (.11,.18) | (.10,.13) | (.10,.11) |
| (.10,.16) | (.10,.17) | (.10,.16) | (.10,.10) | (.14,.22) |
| (.12,.20) | (.10,.18) | (.10,.17) | (.10,.11) | (.10,.15) |
| (.10,.14) | (.10,.10) | (.10,.14) | (.10,.14) | (.14,.20) |
| (.10,.17) | (.10,.14) | (.10,.10) | (.10,.16) | (.10,.17) |

true value of $\delta^*$ is roughly half way between these two estimates. In the table above, we give the estimates obtained in this way when the $\lambda(k)$ are chosen to be i.i.d. and uniformly distributed on [0,1]. The rows correspond to 25 different choices of environment, while the columns give the estimates based on $\alpha_n(k)$ for five consecutive values of $k$. The two components of the ordered pairs are the estimates based on the second and first derivatives respectively, truncated above at .30 and below at .10. Note that only the top third of the environments give estimates consistent with satisfying the conditions of Theorem 5 at $\delta = E\lambda(k)/2 = .25$, while the top two thirds of the environments do at $\delta = E\log(\lambda(k)/2) = .18$. This suggests that

$$E\log(\lambda(k)/2) < \delta^* < E\lambda(k)/2$$

may hold in this case.

## 4. The invariant measures.

In this section, we will sketch the proof of Theorem 6. The assumptions of that theorem are in force throughout the section. For a more detailed proof of the analogous theorem in the homogeneous case, see Theorem 3.13 of Chapter III of Liggett (1985). We will use the "natural" coupling $(\eta_t, \zeta_t)$ of two copies of the contact process which satisfy $\eta_t \leq \zeta_t$ in which the coordinates flip together as much as possible. For configurations $\eta$ and $\zeta$ satisfying $\eta \leq \zeta$, integers $m$ and $n$ satisfying $-\infty \leq m < n \leq \infty$ and a finite set of integers $A$, define $g_{A,m,n}(\eta, \zeta)$ to be the indicator of the set

$$\begin{aligned}\{(\eta, \zeta) :\ &\text{there exist integers } k \text{ and } l \text{ satisfying } m \leq k < l \leq n, (k,l) \supset A \\ &\zeta(k) = \zeta(l) = 1, \eta(k) = \eta(l) = 0, \\ &\eta \equiv \zeta \equiv 1 \text{ on } A, \text{ and } \eta \equiv \zeta \equiv 0 \text{ on } (k,l)\backslash A\}.\end{aligned}$$

Then let

$$f_{m,n}(\eta, \zeta) = \sum_A g_{A,m,n}(\eta, \zeta). \tag{4.1}$$

Note that if the distribution of the coupled process at time t is $\nu$, then $f_{m,n}(\eta_t, \zeta_t)$ decreases by one at a rate which is

$$\geq \sum_{m<j<n} \delta(j) \int g_{\{j\},m,n}(\eta, \zeta) d\nu$$

and $f_{m,n}(\eta_t, \zeta_t)$ increases by one at a rate which is

$$\begin{aligned}&\leq \lambda(m) \int [f_{m-1,n}(\eta, \zeta) - f_{m,n}(\eta, \zeta)]\, d\nu \\ &\quad + \rho(n) \int [f_{m,n+1}(\eta, \zeta) - f_{m,n}(\eta, \zeta)]\, d\nu.\end{aligned}$$

Therefore, if $\nu$ is an invariant measure for the coupled process,

$$\begin{aligned}&\sum_{m<j<n} \delta(j) \int g_{\{j\},m,n} d\nu \\ &\leq \lambda(m) \int [f_{m-1,n} - f_{m,n}] d\nu + \rho(n) \int [f_{m,n+1} - f_{m,n}] d\nu.\end{aligned} \tag{4.2}$$

Since the flip rates are bounded above, the integrands on the right of (4.2) are at most one, and $g_{A,m,n}(\eta,\zeta)$ is increasing in $n$ and decreasing in $m$, it follows that $\nu$ satisfies

$$\sum_j \delta(j) \int g_{\{j\},-\infty,\infty} d\nu < \infty. \tag{4.3}$$

Similarly, if the distribution of the coupled process at time $t$ is $\nu$, then $g_{A,m,n}(\eta_t,\zeta_t)$ decreases by one at a rate which is

$$\leq M|A| \int g_{A,m,n} d\nu,$$

where $M$ is a constant depending only on the bound on the flip rates and $|A|$ denotes the cardinality of $A$, and $g_{A,m,n}(\eta_t,\zeta_t)$ increases by one at a rate which is

$$\geq \sum_{m<j<n} \delta(j) \int g_{A\cup\{j\},m,n} d\nu.$$

Therefore, if $\nu$ is an invariant measure for the coupled process,

$$\sum_{m<j<n} \delta(j) \int g_{A\cup\{j\},m,n} d\nu \leq M|A| \int g_{A,m,n} d\nu. \tag{4.4}$$

Now pass to the limit in (4.4) on $m$ and $n$ as before, use induction on the cardinality of $A$ beginning with (4.3) to conclude that

$$\sum_{|A|=N} \prod_{j\in A} \delta(j) \int g_{A,-\infty,\infty} d\nu < \infty \tag{4.5}$$

for each $N$, if $\nu$ is an invariant measure for the coupled process. When combined with assumption (1.6) on $\delta(j)$ and Hölder's inequality, (4.5) gives

$$\lim_{n\to\infty,m\to-\infty} \frac{1}{n-m} \sum_{|A|=N} \int g_{A,m,n} d\nu = 0. \tag{4.6}$$

Note that by the definition of $g_{A,m,n}$,

$$\sum_A |A|\, g_{A,m,n}(\eta, \zeta) \le n - m \tag{4.7}$$

for all $m, n, \eta$, and $\zeta$. Combining (4.1), (4.6) and (4.7), we conclude that

$$\lim_{n\to\infty, m\to-\infty} \frac{1}{n-m} \int f_{m,n}\, d\nu = 0 \tag{4.8}$$

whenever $\nu$ is an invariant measure for the coupled process. Using this in (4.2), it follows that

$$\int g_{A,-\infty,\infty}\, d\nu = 0 \tag{4.9}$$

for singleton $A$. Using induction on (4.4), it then follows that (4.9) holds for all finite A. Recalling the definition of $g_{A,m,n}$, we see that if $\nu$ is an invariant measure for the coupled process, it must concentrate on

$$\begin{aligned}\big\{(\eta,\zeta) :&\eta \equiv 0, \eta \equiv \zeta,\\ &\text{or there is a } k \text{ so that } \eta \equiv 0 \text{ on } (-\infty, k) \text{ and } \eta \equiv \zeta \text{ on } [k, \infty),\\ &\text{or there is a } k \text{ so that } \eta \equiv \zeta \text{ on } (-\infty, k) \text{ and } \eta \equiv 0 \text{ on } [k, \infty)\big\}.\end{aligned}$$

Therefore, if $\nu$ is an extremal invariant measure for the coupled process, it must concentrate on one of the four sets

$$\{(\eta,\zeta) : \eta \equiv 0\}, \quad \{(\eta,\zeta) : \eta \equiv \zeta\}$$

$$\{(\eta,\zeta) : \text{ there is a } k \text{ so that } \eta \equiv 0 \text{ on } (-\infty, k) \text{ and } \eta \equiv \zeta \text{ on } [k, \infty)\},$$

or

$$\{(\eta,\zeta) : \text{ there is a } k \text{ so that } \eta \equiv \zeta \text{ on } (-\infty, k) \text{ and } \eta \equiv 0 \text{ on } [k, \infty)\}.$$

These properties of extremal invariant measures for the coupled process can be transferred to properties of the original process by using Theorem 2.15 of Chapter III of Liggett (1985). The result is given in the following lemma.

**Lemma 4.10.** *Suppose $\mu_1$ and $\mu_2$ are extremal invariant measures for the contact process which satisfy $\mu_1 \le \mu_2$. Then one of the following must hold:*

(a) $\mu_1 = \mu_2$
(b) $\mu_1$ *= pointmass at* $\eta \equiv 0$.
(c) $\mu_1$ *and* $\mu_2$ *can be coupled so that the coupling measure concentrates on*

$$\{(\eta,\zeta): \text{ there is a } k \text{ so that } \eta \equiv 0 \text{ on } (-\infty,k) \text{ and } \eta \equiv \zeta \text{ on } [k,\infty)\}.$$

(d) $\mu_1$ *and* $\mu_2$ *can be coupled so that the coupling measure concentrates on*

$$\{(\eta,\zeta): \text{ there is a } k \text{ so that } \eta \equiv 0 \text{ on } (-\infty,k) \text{ and } \eta \equiv 0 \text{ on } [k,\infty)\}.$$

*Proof of Theorem 6.* Suppose that $\nu$ is an extremal invariant measure which is different from $\nu_0$ and $\nu_1$. Applying Lemma 4.10 with $\mu_1 = \nu$ and $\mu_2 = \nu_1$, we conclude that the pair $\nu$ and $\nu_1$ satisfy (*c*) or (*d*) of that lemma. By symmetry, we may assume that they satisfy (*c*). There is at most one extremal invariant measure which can be coupled with $\nu_1$ in that way, since if there were two, then those two could be coupled together in such a way that they differed at only finitely many sites. But then there would be an extremal invariant measure for the coupled process which concentrates on

$$\{(\eta,\zeta): \eta(k) = \zeta(k) \text{ for all but finitely many } k \text{ but } \eta(k) \ne \zeta(k) \text{ for some } k\}.$$

But this is impossible, since all such pairs $(\eta,\zeta)$ lead at a positive rate to the absorbing class $\{(\eta,\zeta): \eta \equiv \zeta\}$.

## References

[1] M. Bramson, R. Durrett and R. Schonmann, *The contact process in a random environment*, Ann. Probab. (1991).

[2] H. Brezis and A. Pazy, *Convergence and approximation of semigroups of nonlinear operators in Banach spaces*, J. Funct. Anal **9** (1972), pp. 63–74.

[3] T. Brox, *A one-dimensional diffusion process in a Wiener medium*, Ann. Probab. **14** (1986), pp. 1206–1218.

[4] R. Durrett, *Multidimensional random walks in a random environment with subclassical limiting behavior*, Comm. Math. Phys. **104** (1986), pp. 87–102.

[5] R. Durrett, "Lecture Notes on Particle Systems and Percolation," Wadsworth, Pacific Grove, 1988.

[6] R. Durrett, *A new method for proving existence of phase transitions*, This volume (1990).

[7] I. M. Ferreira, *The probability of survival for the biased voter model in a random environment*, Stoch. Proc. Appl. **34** (1990), pp. 25–38.

[8] A. Greven, *The coupled branching process in random environment*, Ann. Probab. **13** (1985), pp. 1133–1147.

[9] D. Griffeath, *Additive and Cancellative Interacting Particle Systems*, Springer Lecture Notes in Mathematics **724** (1979).

[10] D. Griffeath, *The basic contact process*, Stoch. Proc. Appl. **11** (1981), pp. 151–186.

[11] T. E. Harris, *A lower bound for the critical probability in a certain percolation process*, Proc. Camb. Phil. Soc. **56** (1960), pp. 13–20.

[12] T. E. Harris, *Contact interactions on a lattice*, Ann. Probab. **2** (1974), pp. 969–988.

[13] T. E. Harris, *On a class of set-valued Markov processes*, Ann. Probab. **4** (1976), pp. 175–194.

[14] T. E. Harris, *A correlation inequality for Markov processes in partially ordered state spaces*, Ann. Probab. **5** (1977), pp. 451–454.

[15] T. E. Harris, *Additive set-valued Markov processes and graphical methods*, Ann. Probab. **6** (1978), pp. 355–378.

[16] R. Holley and T. M. Liggett, *The survival of contact processes*, Ann. Probab. **6** (1978), pp. 198–206.

[17] H. Kesten, M. Kozlov and F. Spitzer, *A limit law for random walk in a random environment*, Compos. Math. **30** (1975), pp. 145–168.

[18] T. M. Liggett, "Interacting Paricle Systems," Springer, New York, 1985.

[19] N. H. Pavel, *Nonlinear Evolution Operators and Semigroups*, Springer Lecture Notes in Mathematics **1260** (1987).

[20] Y. G. Sinai, *The limiting behavior of a one-dimensional random walk in a random environment*, Th. Probab. Appl. **27** (1982), pp. 256–268.

[21] F. Solomon, *Random walks in a random environment*, Ann. Probab. **3** (1975), pp. 1–31.

Research supported in part by NSF Grant DMS 86-01800

Department of Mathematics
University of California, Los Angeles
Los Angeles, CA 90024

# A New Method for Proving the Existence of Phase Transitions

Rick Durrett

In this paper we will describe a new method for proving the existence of phase transitions, developed in joint work with Maury Bramson, and apply it to four examples. The material here is related to Ted Harris' work in at least four ways. First, all four examples are related to the contact process, a system he invented in 1974. Second, the method we are going to describe was used in his 1974 paper to show that the contact process has a phase transition. Third, all of our processes are constructed from Poisson processes using either the procedure he invented in 1972 for general finite range models or the graphical representations he invented for the contact process and related systems in 1978. Finally, branching processes play an important role in the proofs below.

We begin by describing the four examples we are going to treat. We will keep the discussion to a minimum to allow the parallels to stand out. At the beginning of the references we will explain who proved what and where the papers will appear.

*Example 1.* The *contact process* is a Markov process whose state at time $t$ is $\xi_t \subset \mathbb{Z}^d$ and which evolves according to the following rules:

(i) Particles die at rate $\delta$, i.e., if $x \in \xi_t$ then $P(x \notin \xi_{t+s}|\xi_t) \sim \delta s$ as $s \to 0$.

(ii) If $x$ is occupied and $y$ is a neighbor of $x$ (i.e. $|x-y| = 1$), births occur from $x$ to $y$ at rate 1. Here, and below, $|z| = |z_1| + ... + |z_d|$.

(iii) If $y$ is occupied the two particles coalesce to 1, or if you prefer, the birth is suppressed. Either way you look at it, there can be at most one particle per site.

*Note.* It is customary to make the death rate 1 and the birth rate $\lambda(= 1/\delta)$. The reader will see the reason for deviating from the usual practice as the story unfolds.

Let $\xi_t^A$ denote the system starting with $\xi_0^A = A$ and write $\xi_t^0$ for $\xi_t^{\{0\}}$. Let

$$\delta_c = \sup\{\delta : P(\xi_t^0 \neq \phi \text{ for all } t) > 0\}.$$

The first thing to show is that there is a phase transition.

**Theorem 1A.** $0 < \delta_c \leq 2d$.

The upper bound is trivial. Each particle gives birth to a new one at rate $\leq 2d$ and dies at rate $\delta$, so if $\delta > 2d$ comparing with a branching process shows that the system *dies out*, i.e., $P(\xi_t^0 \neq \emptyset \text{ for all } t \geq 0) = 0$. The argument that allows us to prove Theorem 1A also allows us to prove a convergence theorem. Let $\delta_\phi$ be the pointmass on the empty set and let $\xi_\infty^1$ denote the limit starting from $\xi_0^1 = \mathbb{Z}^d$. (This exists. See Chapter 4 in Durrett (1988).)

**Theorem 1B.** *Let $\tau^A = \inf\{t : \xi_t^A = \phi\}$. If $\delta$ is small then*

$$\xi_t^A \Rightarrow \delta_\phi P\left(\tau^A < \infty\right) + \xi_\infty^1 P\left(\tau^A = \infty\right).$$

Here $\Rightarrow$ denotes weak convergence, which in this setting is convergence of finite dimensional distributions. A consequence of Theorem 1B is that for small $\delta$, all stationary distributions are a convex combination of $\delta_\phi$ and $\xi_\infty^1$. Similar remarks apply to Theorems 2B–4B.

*Example 2.* The *annihilating branching process* is a Markov process whose state at time $t$ is $\xi_t \subset \mathbb{Z}^d$, and which evolves according to the following rules:

(i) Particles die at rate $\delta \geq 0$.
(ii) If $x$ is occupied and $y$ is a neighbor of $x$ (i.e. $|x-y| = 1$), births occur from $x$ to $y$ at rate 1.
(iii) If $y$ is occupied then the two particles annihilate and an empty site results.

Rules (i) and (ii) are like those of the contact process. In (iii) we have replaced coalesce by annihilate. Let $\xi_t^A$ be the process with $\xi_0^A = A$ and write $\xi_t^0$ for $\xi_t^{\{0\}}$. Let

$$\delta_c = \sup\{\delta : P(\xi_t^0 \neq \phi \text{ for all } t) > 0\}.$$

**Theorem 2A.** $0 < \delta_c \leq 2d$.

The upper bound follows from the argument used for Example 1. Again as a byproduct of the proof of Theorem 2A we get a convergence theorem. Let $\nu_{1/2}$ be the product measure with density 1/2 (i.e., $x \in \xi_0$ are independent and have probability 1/2). Let $\xi_\infty^{1/2}$ denote the limit starting from $\nu_{1/2}$. (This exists. See Chapter 5 of Durrett (1988).)

**Theorem 2B.** *Let* $\tau^A = \inf\{t : \xi_t^A = \phi\}$. *If* $\delta$ *is small then*

$$\xi_t^A \Rightarrow \delta_\phi P\left(\tau^A < \infty\right) + \xi_\infty^{1/2} P\left(\tau^A = \infty\right).$$

The result is interesting even for $\delta = 0$. In that case, two special things happen. First, $\nu_{1/2}$ is a stationary distribution., i.e., $\xi_\infty^{1/2} = \nu_{1/2}$. Second, particles only die by being murdered, so $P(\tau^A = \infty) = 1$ for $A \neq \phi$, and in this case the conclusion of Theorem 2B can be written as: If $A \neq \phi$ then $\xi_t^A \Rightarrow \nu_{1/2}$.

*Example 3.* The *long range contact process* is a Markov process whose state at time $t$ is $\xi_t \subset \mathbb{Z}^d/M = \{z/M : z \in \mathbb{Z}^d\}$ where $M$ is a large integer. The system evolves according to the following rules:

(i) Particles die at rate 1 and give birth at rate $\beta$.
(ii) A particle born at $x$ is sent to a point $y$ chosen at random from $\{y : \|x - y\|_\infty \leq 1\}$ where $\|z\|_\infty = \sup |z_i|$.
(iii) If $y$ is occupied then the two particles coalesce to 1.

The motivation for this definition can be seen in the other name for this process: "crab grass". It indicates we are thinking of a lattice in which the spaces between the sites are small compared to the distance offspring are displaced from their parents. This interpretation begs for (ii) to be generalized, but the reader will see from the proof that this can be done easily.

Let $\xi_\infty^A$ denote the process with $\xi_0^A = A$ and write $\xi_t^0$ for $\xi_t^{\{0\}}$. Let

$$\beta_c = \inf\{\beta : P(\xi_t^0 \neq \phi \text{ for all } t) > 0\}.$$

The argument used for the first two examples shows that $\beta_c \geq 1$. When $M \to \infty$, the contact process approaches a branching random walk $Z_t \subset \mathbf{R}$ in which

(i) Particles die at rate 1 and give birth at rate $\beta$.
(ii) A particle born at $x$ is sent to a point $y$ chosen at random from $\{y : \|x - y\|_\infty \le 1\}$ where $\|z\|_\infty = \sup |z_i|$.

The branching random walk has $\beta_c = 1$. The next result says that the critical value and survival probability are continuous functions of $M$.

**Theorem 3A.** *As $M \to \infty$, $\beta_c(M) \to 1$ and if $\beta > 1$*

$$P(\xi_t^0 \neq \phi \text{ for all } t) \to (\beta - 1)/\beta.$$

Here, as the remark before the theorem suggests, the right hand side is the survival probability for the branching random walk. Bramson, Durrett, and Swindle (1989) proved the more precise result

$$\beta_c(M) - 1 \approx \begin{cases} 1/M^{2/3} & d = 1 \\ (\log M)/M^2 & d = 2 \\ 1/M^d & d \ge 3. \end{cases}$$

where $\approx$ means that if we multiply the right hand side by a small (large) constant then we get a lower (an upper) bound. The proof of the last result is rather involved. (Even the referee couldn't bear to read more than half of the paper.) In Section 4 we will give a simple proof of Theorem 3A. As in the first two cases, our proof gives a convergence theorem.

**Theorem 3B.** *Let $\beta > 1$. If $M > M_\beta$ then*

$$\xi_t^A \Rightarrow \delta_\phi P\left(\tau^A < \infty\right) + \xi_\infty^1 P\left(\tau^A = \infty\right).$$

Here $\xi_\infty^1$ is the limit starting from $\xi_0^1 = \mathbb{Z}^d$, which exists by Chapter 4 of Durrett (1988).

*Example 4.* The *multitype contact process* is a Markov process whose state at time $t$ is $\xi_t : \mathbb{Z}^d/M \to \{0, 1, 2\}$. We think of 0 = grass, 1 = bushes, and 2 = trees, and formulate the evolution as follows:

(i) 1's and 2's each die (i.e., become 0) at rate 1. 1's (2's) give birth at rate $\lambda_1$ (resp. $\lambda_2$).
(ii) If the birth occurs at $x$ the offspring is sent to a site chosen at random from $\{y : \|y - x\|_\infty \le 1\}$.
(iii) If $\xi_t(y) \ge \xi_t(x)$ then the birth is suppressed. If $\xi_t(y) < \xi_t(x)$ then site $y$ changes its state to $\xi_t(x)$.

The last rule reflects the fact that grass, bushes, trees is a successional sequence, i.e., each plant can displace its predecessor. Since 2's can displace 1's or 0's, it should be clear that $\zeta_t = \{y : \xi_t(y) = 2\}$ is Example 3.

It is not hard to show that in this model if $\lambda_2$ is too large then the *1's die out*, that is, if $|\zeta_t| = |\{y : \xi_t(y) = 2\}| = \infty$, then $\xi_t \Rightarrow \mu_2$ the limit starting from $\xi_0^2(x) \equiv 2$. When $d = 1$ and $M = 1$, this holds whenever $\lambda_2 > \lambda_c$. As the last observation may suggest, it is more delicate to prove that the other alternative can occur.

**Theorem 4A.** *If $\lambda_1 > \lambda_2^2 > 1$ then coexistence occurs for large $M$. That is, there is a translation invariant measure $\mu_{12}$ with $\mu_{12}\left(\Lambda_\infty^1 \cap \Lambda_\infty^2\right) = 1$ where $\Lambda_\infty^i = \{\eta : \eta(x) = i$ for infinitely many $x\}$.*

To explain the condition in the theorem, we observe that an extension of the results quoted for Example 3 shows that if $\lambda_2 > 1$ and $M$ is large, $\mu_2$ is almost a product measure in which 2's have density $(\lambda_2 - 1)/\lambda_2$. Setting $M = \infty$, we define a *mean field version* of the set of sites occupied by 1's, $Z_t$, which evolves as follows:

(i) Each particle (i.e., element of $Z_t$) dies at rate 1 and gives birth at rate $\lambda_1$.

(ii) The offspring of a particle at $x$ is sent to a point $y$ chosen at random from $\{y : \|x-y\|_\infty \le 1\}$. We flip a coin with probability $(\lambda_2 - 1)/\lambda_2$ of heads to see if the site is occupied by a 2 or not. If we decide the site is occupied by a 2 the birth is suppressed. (We will be interested in what happens when $Z_0$ is finite, so we can ignore the possibility of hitting a 1.)

(iii) To simulate births from sites occupied by 2's, for each $x \in Z_t$ at times of a Poisson process with rate $\lambda_2$ we pick a site at random from $\{y : \|x-y\|_\infty \le 1\}$ and flip another coin to see if $y$ is occupied by a 2. If it is we remove $x$ from $Z_t$.

$Z_t$ is just a branching process in which particles die at rate $1 + \lambda_2 \times (\lambda_2 - 1)/\lambda_2 = \lambda_2$ and are born at rate $\lambda_1 \cdot 1/\lambda_2$, so in order for the 1's to survive we need $\lambda_1/\lambda_2 > \lambda_2$. The careful reader will have noticed that we just argued that the condition should be necessary, whereas our theorem says it is sufficient. However, we do not know how to prove the other direction.

**Conjecture.** *If $\lambda_1 < \lambda_2^2$ then the 1's die out for large $M$.*

As in the first three cases, our proof of Theorem 4A leads to a convergence theorem. We call it a conjecture to indicate the fact that a proof has not yet been written up (down?). Let $\nu_i$ be the limit starting from $\xi_0^i(x) \equiv i$.

**Conjecture 4B.** *Let $\tau^i = \inf\{t : \xi_t \in \Lambda_0^i\}$ where $\Lambda_0^i = \{\eta : \eta(x) \neq i$ for all $x\}$. If $\lambda_1 > \lambda_2^2 > 1$ and $M$ is large, then*

$$\xi_t \Rightarrow \mu_0 P\left(\tau^1 < \infty, \tau^2 < \infty\right) + \mu_1 P\left(\tau^1 = \infty, \tau^2 < \infty\right) + \mu_2 P\left(\tau^1 < \infty, \tau^2 = \infty\right) + \mu_{12} P\left(\tau^1 = \infty, \tau^2 = \infty\right).$$

The rest of this paper is devoted to a discussion of the proofs of Theorem 1A-4A. To keep focused on the problem of the existence of phase transitions and make the workings of our method stand out as clearly as possible, we will ignore Theorems 1B-4B. In each case the proof of the convergence theorem uses the fact that the process dominates supercritical oriented site percolation but relies on a "duality equation" specific to the process. Remarks in the references indicate where the proofs can be found in the literature.

## Section 2. The Big Picture

The method of proof for Theorems 1A-4A (in what follows we will drop the A) is the same. A cocktail party (or ten second sound bite) summary is

*We prove our results by showing that when viewed on suitable length and time scales, the system dominates 1-dependent oriented percolation with p close to 1.*

The first step then is to define 1-dependent oriented percolation. Let

$$\mathcal{L} = \{(m,n) \in \mathbb{Z}^2 : m+n \text{ is even}\}.$$

Given random variables $\omega(m,n), (m,n) \in \mathcal{L}$, that indicate whether $(m,n)$ is open (1) or closed (0), we say that $(y,n)$ can be reached from $(x,m)$ and write $(x,m) \to (y,n)$ if there is a sequence of points $x_m = x, \ldots, x_n = y$ so that $|x_\ell - x_{\ell+1}| = 1$ for $m \leq \ell < n$, and $\omega(x_\ell, \ell) = 1$ for $m \leq \ell \leq n$. Let $C = \{z : (0,0) \to z\}$ be the *cluster containing* (0,0) and let $\Omega_\infty = \{|C| = \infty\}$ be the event that *percolation occurs.*

The last paragraph explains the "oriented percolation". To explain the rest of the name, let $\|(m,n)\|_L = (|m|+|n|)/2$. We divide by two so that the distance from $(m,n)$ to $(m+1,n+1)$ is 1. If we have

$$(*)\quad P\left(\omega\left(z^i\right)=0 \text{ for } i=1,\dots,k\right) \le \varepsilon^k \text{when } z^1,\dots,z^k \in L \text{ have } \left\|z^i - z^j\right\|_L > 1$$

we say that $\omega$ is a *1-dependent system with density* $\geq 1-\varepsilon$. The last assumption is exactly what we need to use the arguments in Section 10 of Durrett (1984) to conclude that

$$P(|C|<\infty) \le \sum_{n=4}^{\infty} 3^n \varepsilon^{n/36}.$$

We have nothing to add to the argument given there so we will not repeat it. Summing the last series gives

(1) If $(*)$ holds and $\varepsilon \le 6^{-36}$ then $P(|C|<\infty) \le 2\cdot 3^4 \cdot \varepsilon^{1/9}$.

The right hand side of the inequality is $\le 2^{-3} < 1$, so we have shown that percolation occurs when $\varepsilon$ is ridiculously small. For some of our applications we will need a result dual to (1).

(2) Let $W_n = \{y : (x,0) \to (y,n) \text{ for some } x \in 2\mathbb{Z}\}$. If $(*)$ holds and $\varepsilon \le 6^{-36}$ then

$$P(y \notin W_n) \le 2\cdot 3^4 \cdot \varepsilon^{1/9} \text{ whenever } (y,n) \in \mathcal{L}.$$

*Proof.* The variables $\bar{\omega}(j,k) = \omega(j+m, n-k)$ satisfy $(*)$. If there is a path from (0,0) to $(\cdot, n)$ in the $\bar{\omega}$ system then the desired event occurs in the $\omega$ system.

Before entering into a general discussion of our method of proof we need to make a definition. In each case the system we are interested in is a perturbation of a simple system. In Examples 1 and 2 the *simple system* is the process with $\delta = 0$. In Examples 3 and 4 it is the process with $M = \infty$, i.e., the branching random walk in Example 3 and the mean field version in Example 4.

To relate the four systems defined in the introduction to oriented percolation we will use the following steps:

I. We will construct the process from a family of Poisson processes (a.k.a. graphical representation). This will enable us to construct on one probability space copies of the process starting from any configuration at any time.

II. We let $e_1 = (1, 0, ...0)$, $\varphi(m,n) = (2mLe_1, nT)$ for $(m,n) \in \mathcal{L}$, and $B_{m,n} = \varphi(m,n) + B$, where the box $B \subset \mathbb{R}^{d+1}$ is chosen so that $B_{j,k} \cap B_{m,n} = \phi$ when $\|(j,k) - (m,n)\|_{\mathcal{L}} > 1$, we set $\omega(m,n) = 1$ if a certain good event happens in the graphical representation in $B_{m,n}$.

The good event is chosen so that (1) or (2) above gives us the desired result. Properties of the Poisson process imply that our events are 1-dependent. Last but not least, we show that if $\varepsilon > 0$ it is possible to pick $L$ and $T$ so that the good event has probability $> 1 - \varepsilon$ for the simple system.

III. With our attention fixed on a finite box, it follows from "continuity" that if $\delta$ is small in Examples 1 and 2 or $M$ is large in Examples 3 and 4, then the good event has probabilty $> 1 - 2\varepsilon$. If $\varepsilon \leq 6^{-36}/2$, we have (1) and (2) and the desired result follows.

Before entering into the details of the four applications we would like to observe that the hard work (step II) is done for the simple system and then "continuity" extends the conclusion to the system we want to study. A consequence of this approach is that once we have shown that the simple system dominates oriented percolation, it follows that this is also true for a small perturbation by a translation invariant mechanism with bounded rates. E.g., in Example 2 we can, instead of adding spontaneous deaths, change the rules of the simple process so that a particle at $x$ jumps to $y$ at rate $\delta p(x,y)$, where $p(x,y) = f(x-y)$, and if y is occupied the two particles annihilate. In this way the proof of Theorem 2 generalizes immediately to give a result of Bramson and Gray (1985). They proved their result for simple random walk, $f(z) = 1/2d$ when $|z| = 1$, by the "contour method" and only wrote out the details for $d = 1$. To really appreciate the virtues of our new approach, the reader should try to use the contour method to prove Theorem 2 or extend the result of Bramson and Gray to a general random walk.

Climbing down from our "we're better than the contour method" soapbox and returning to the business of educating the reader: the four examples are treated in order of increasing complexity: 1, 3, 4, 2. We have done the proofs in parallel as much as possible to bring out the similarities in the arguments. We felt that it was important to give complete proofs to show how the method *really* works, but to quote one of our favorite jokes: "The method is like cutting a tomato with a chainsaw. There is more than enough power to do the job, but the result is somewhat messy." By the end of Section 4, the reader will understand the true meaning of the last quote—the method requires a number of tedious details. In reading the proofs we hope that the reader will concentrate on how the method works as enunciated in I–III above and ignore the mountain of mindless computations needed to carry it out.

## 3. Proof of Theorem 1

We begin by constructing the process from a collection of Poisson processes. For $x, y \in \mathbb{Z}^d$ with $|x-y| = 1$, let $\{T_n^{x,y} : n \geq 1\}$ and $\{U_n^x : n \geq 1\}$ be the arrival times of Poisson processes with rates 1 and $\delta$. As the reader can probably guess from the rates: at times $U_n^x$ we kill the particle at $x$ if one is present, and at time $T_n^{x,y}$ there is a birth from $x$ to $y$ if $x$ is occupied.

Even though there are infinitely many Poisson processes and hence no first arrival, an idea of Harris (1972) allows us to construct the process starting from any $\xi_0$. Consider a random graph in which $x$ and $y$ are connected if $\min(T_1^{x,y}, T_1^{y,x}) \leq \tau$. If $\tau$ is chosen small enough, the probability of connection is $< 1/2d$ and a simple argument (compare with a branching process) shows that all the components of our random graph are finite. The evolution of each component is unaffected by the others and can be computed separately. In this way we can construct the process up to time $\tau$, and iterating constructs the process for all time.

A simple comparison shows that for fixed $\delta, \xi_t^0$ is larger in $d$ dimensions than in $d = 1$ (i.e., the two processes can be constructed on the same space so that one process contains the other). This shows $\delta_c(\mathbb{Z}^d) \geq \delta_c(\mathbb{Z})$ so it suffices to prove the result when $d = 1$. Let

$$\mathcal{L} = \{(m,n) \in \mathbb{Z}^2 : m+n \text{ is even}\}$$

$$\varphi(m,n) = (2mL, nT), \quad \text{where } T = \kappa L$$

$$B = (-4L, 4L) \times [0,T] \quad B_{m,n} = \varphi(m,n) + B$$

$$I = [-L, L] \quad I_m = 2mL + I.$$

$\kappa$ is a constant that will be chosen later. We will say that $(m,n) \in \mathcal{L}$ is occupied if $\xi_{nT} \supset I_m$. Our aim is to show:

(0) Let $\varepsilon > 0$. If $L \geq L_\varepsilon$ then the process of occupied sites dominates 1-dependent oriented site percolation with $p = 1 - \varepsilon$.

Consider the process with $\delta = 0$ and suppose $\xi_0 = [-L, L]$. Since there are no deaths, the state at any time will be an interval $[\ell_t, r_t]$ and the end points will move as follows:

$$r_t \to r_t + 1 \text{ at rate } 1 \text{ and } \ell_t \to \ell_t - 1 \text{ at rate } 1.$$

The strong law of large numbers implies that

$$L^{-1} \sup_{0 \le s \le L} (r_s - r_0 - s) \to 0 \ a.s. \text{ as } L \to \infty,$$

Let $\kappa = 5/2$ and $T = \kappa L$. We let $\omega(m,n) = 1$ if $\zeta_t^{m,n}$, the process starting with $I_m$ occupied at time $nT$, contains $I_{m-1}$ and $I_{m+1}$ at time $(n+1)T$ and has $\zeta_t^{m,n} \times \{t\} \subset B_{m,n}$ for all $t \in [nT,(n+1)T)$. The events are defined in such a way that $\omega(j,k)$ and $\omega(m,n)$ are independent if $\|(j,k)-(m,n)\|_{\mathcal{L}} > 1$ and if $(0,0) \to (m,n)$ in the percolation process on $\mathcal{L}$ then

$$\xi_{nT}^{I(0)} \supset I(m)$$

(1) implies that if $L \ge L_\varepsilon, P(\omega(m,n)=1) > 1-\varepsilon$, so $\omega$ dominates a 1-dependent system with density $\ge 1-\varepsilon$.

To extend the last conclusion to $\delta > 0$, let $\mathcal{D} = \{(x, U_n^x) : n \ge 1, x \in \mathbb{Z}^d\}$ be the set of deaths in the graphical representation and let $\omega'(m,n) = 1$ if $\omega(m,n) = 1$ and $\mathcal{D} \cap B_{m,n} = \phi$. If we $L = L_\varepsilon$ then for $\delta \le \delta_\varepsilon$, $\omega'$ dominates a 1-dependent system with density $\ge 1-2\varepsilon$. The desired result now follows from (2.1).

The proof above generalizes immediately to solve two other problems. We continue the numbering in the introduction.

*Example 5.* The *centipede* is a Markov process whose state at time $t$ is $\xi_t \subset \mathbb{Z}$ and which evolves according to the following rules:

(i) Particles die at rate $\delta$.
(ii) If $x+1,...,x+M$ are occupied then births occur at $x$ and $x+M+1$ at rate 1.
(iii) If a birth occurs at an occupied site then the two particles coalesce to 1.

The name refers to the fact that we are thinking about the case $M = 100$.

The case $M = 2$ has been studied under the name *sexual reproduction.* Let $\xi_t^A$ denote the process starting with $\xi_0^A = A$ and write $\xi_t^*$ for $\xi_t^{\{1,\cdots,M\}}$. Let

$$\delta_c = \sup\{\delta : P(\xi_t^* \ne \phi \text{ for all } t) > 0\}.$$

**Theorem 5.** $0 < \delta_c \leq 2/M$.

The upper bond comes from observing that when there are $k$ particles the total death rate is $k$ and the maximum birth rate is $\leq 2k/M$. The proof of the lower bound is almost the same as the one given for the contact process. If $\delta = 0$ and $r_0 - \ell_0 \geq M$ then $r_t$ and $\ell_t$ behave as indicated above, so the proof of Theorem 1 works if we pick $L \geq M$.

*Example 6.* The *stability theorem of Gray and Griffeath.* Consider a one dimensional nearest neighbor spin system on $\mathbb{Z}$ with birth rates $\beta_{ij}$ and death rates $\delta_{ij}$. That is, $\xi_t$ is a Markov process with state space $\{0,1\}^{\mathbb{Z}}$ which has

$$P\left(\xi_{t+s}(x) = 1 | \xi_t(x-1) = i,\ \xi_t(x) = 0,\ \xi_t(x+1) = j\right) \sim \beta_{ij} s,$$
$$P\left(\xi_{t+s}(x) = 0 | \xi_t(x-1) = i,\ \xi_t(x) = 1,\ \xi_t(x+1) = j\right) \sim \delta_{ij} s,$$

as $s \to 0$. Gray and Griffeath (1982) investigated the class of models that have

$$\beta_{00} = 0 \quad \beta_{11} \geq \beta_{01}, \beta_{10} \quad \delta_{11} = 0 \quad \delta_{00} \geq \delta_{01}, \delta_{10}.$$

The inequalities say that the system is attractive, and imply that if $\xi_0(x) \leq \bar{\xi}_0(x)$, then copies of the process starting from these initial configurations can be constructed so that $\xi_0(x) \leq \bar{\xi}_0(x)$ for all $x$ and $t$. The equalities $\beta_{00} = 0$ and $\delta_{11} = 0$ imply that the states $\underline{0}$ = "all 0's" and $\underline{1}$ = "all 1's" are absorbing states.

Gray and Griffeath investigated the stability of the fixed point $\underline{1}$ under the perturbation $\beta^\theta_{ij} = \beta_{ij}, \delta^\theta_{ij} = \delta_{ij} + \theta$. The perturbed flip rates are attractive, so if we start the perturbed system from $\xi^\theta_0(x) \equiv 1$ then $\xi^\theta_t \Rightarrow \nu^\theta_1$ where $\nu^\theta_1$ is a translation invariant stationary distribution. (See Liggett (1985), p. 135.) If $\nu^\theta_1\{\xi : \xi(x) = 1\} \uparrow 1$ as $\theta \downarrow 0$ (the last probability does not depend on $x$), then we say $\underline{1}$ is stable. If $\nu^\theta_1\{\xi : \xi(x) = 1\} = 0$ for all $\theta > 0$, then we say $\underline{1}$ is unstable. Although it is not immediately clear from the definition that something which is not unstable is stable, the next result, due to Gray and Griffeath (1982), shows this is true and identifies when the two cases occur.

**Theorem 6.**

*If* $\beta_{01} + \beta_{10} \leq \delta_{01} + \delta_{10}$, *then* $\underline{1}$ *is unstable*

*If* $\beta_{01} + \beta_{10} > \delta_{01} + \delta_{10}$, *then* $\underline{1}$ *is stable*

The first conclusion in the theorem is easy to prove and is left as an exercise for the reader. A proof of the second conclusion for the special case $\beta_{01} = \beta_{10} = \beta > \delta = \delta_{01} = \delta_{10}$ can be obtained from the proof of Theorem 1A. (For the proof of the full result by this method see Bramson and Durrett (1989). Consider the process with $\theta = 0$ and suppose $\xi_0 = [-L, L]$. Since $\beta_{00} = 0$ and $\delta_{11} = 0$, the state at any time will be an interval $[\ell_t, r_t]$. When $r_t - \ell_t \geq 1$ the endpoints move as follows:

$$r_t \to \begin{cases} r_t + 1 & \text{at rate } \beta \\ r_t - 1 & \text{at rate } \delta \end{cases} \quad \text{and } \ell_t \to \begin{cases} r_t - 1 & \text{at rate } \beta \\ r_t + 1 & \text{at rate } \delta. \end{cases}$$

Let $c = \beta - \delta$, $\kappa = 5L/2c$, and $T = \kappa L$. Using the strong law of large numbers as in the proof of Theorem 1, we see that if $L$ is large and we start with $I_0 = [-L, L]$ occupied at time 0 then with high probability $\xi_T$ will contain $I_1$ and $I_{-1}$, and for all $t \leq T$, $\xi_t$ will be contained in $(-4L, 4L)$. With this established the rest of the proof is the same as before.

## 4. Proof of Theorem 3

We treat Theorem 3 next because its proof is the second simplest. Let $\mathcal{L} = \{(m, m) \in \mathbb{Z}^2 : m + n \text{ is even}\}$, $e_1 = (1, 0, \ldots 0)$, and

$$\varphi(m, n) = (2mLe_1, nT) \quad \text{where } T = L^2$$
$$B = (-2L, 2L)^d \times [0, L^2] \quad B_{m,n} = \varphi(m, n) + B$$
$$I = [-L, L]^d \quad I_m = 2mLe_1 + I.$$

Notice that $B_{m,n} \cap B_{m,n+2} = \phi$. We will use this choice of $B$ in the next two examples. We will say that $(m, n) \in \mathcal{L}$ is occupied if $|\xi_{nT} \cap I_m| \geq K$. Our aim will be to show:

(0) Let $\varepsilon > 0$. If $K, L$ are chosen appropriately then the process of occupied sites dominates independent oriented site percolation with $p = 1 - \varepsilon$.

**A. Preliminaries.**

We begin by proving the analogous result for a branching random walk $Z_t$ in which:

(i) Particles die at rate 1 and give birth at rate $\beta$.

(ii) A particle born at $x$ is sent to a point y chosen at random from $\{y : \|x - y\|_\infty \leq 1\}$.

Let $Z_t^x$ denote the system with $Z_0^x = \{x\}$. To have events that are independent we consider $\bar{Z}_t^x$ a modification of $Z_t^x$ in which particles that land outside $[-2L+1, 2L-1]^d$ are killed. The $+1$ and $-1$ are to leave room for the limit $M \to \infty$. Let $m(t,x,A) = E|\bar{Z}_t^x \cap A|$. It is easy to check that

$$m(t,x,A) = e^{(\beta-1)t} P(\bar{S}_t^x \in A) \tag{1}$$

where $\bar{S}_t^x$ is a random walk that starts at $x$, takes steps at rate $\beta$ and is killed when it lands outside $[-2L+1, 2L-1]^d$. (Observe that both sides of (1) satisfy the same differential equation.) Let $I_1' = 2Le_1 + [-L+1, L-1]^d$, i.e. $I_1$ shrunk by a little bit. Donsker's theorem implies that if $T = L^2$ and $x/L \to \theta \in [-1,1]^d$

$$P(\bar{S}_t^x \in I_1') \to \psi(\theta) \tag{2}$$

where $\psi(\theta) = P_\theta(B_t \in [-2,2]^d \text{ for } t \le 1, B_1 \in 2e_1 + [-1,1]^d)$ and $B_t$ is a constant multiple of $d$-dimensional Brownian motion. $\psi(\theta) > 0$ and continuous, so a simple argument (suppose not and extract a convergent subsequence) shows

$$\liminf_{L\to\infty} \left[ \inf_{x\in[-L,L]^d} P\left(\bar{S}_t^x \in I_1'\right)\right] \ge \inf_{\theta\in[-1,1]^d} \psi(\theta) > 0. \tag{3}$$

It follows from (2) and (3) that we can pick $L$ large enough so that

$$\inf_{x\in[-L,L]^d} E\left|\bar{Z}_T^x \cap I_1'\right| \ge 2. \tag{4}$$

Let $\bar{Z}_t^A$ denote the modified branching process with $\bar{Z}_t^A = A$. (4) implies

$$E\left|\bar{Z}_T^A \cap I_1'\right| \ge 2|A| \tag{5}$$

$$E\left|\bar{Z}_T^x \cap I_1'\right|^2 \le E\left|Z_T^x\right| = C_T < \infty \tag{6}$$

so it follows from Chebyshev's inequality that if $A \subset [-L,L]^d$ has $|A| \ge K$ then

$$P\left(\left|\bar{Z}_T^A \cap I_1'\right| < K\right) \le var\left(\left|\bar{Z}_T^A \cap I_1'\right|\right) / (2|A| - K) \le C_T/K. \tag{7}$$

From the last result it follows that

(8) If $\varepsilon > 0$ and $K$ is large then
$$P\left(\left|\bar{Z}_T^A \cap I_1'\right| < K\right) \leq \varepsilon \text{ for any } A \subset [-L, L]^d \text{ with} |A| \geq K.$$

**B. Continuity Argument.**

(8) shows that if $A \subset I_0$ has $|A| \geq K$ then with high probability $\overline{Z}_T^A$ will have at least $K$ particles in $I_{-1}$ and in $I_1$. The next step is to prove the corresponding result for the contact process. Let $\overline{\xi}_t^A$ be a modification of the contact process with $\overline{\xi}_0^A = A$ in which births outside $(-2L, 2L)^d$ are not allowed. We begin by observing that the number of points in

$$V_t = \bigcup_{0 \leq s \leq t} \overline{\xi}_s^A$$

is dominated by a branching process $W_t$ in which births occur at rate $\beta$ and deaths occur at rate 0. If $|A| < 2K$ then $EW_t \leq 2K \exp(\beta t) < \infty$, so $P\left(V_t > M^{1/3}\right) \to 0$. When $V_t \leq M^{1/3}$ the probability of having a birth land on an occupied site is

(9) $$\leq M^{1/3}\left(M^{1/3}/\left(2M+1\right)^d\right) \to 0.$$

To deal with the spatial location of particles, we observe that given a realization of the branching random walk $\overline{Z}_s^A$ we can construct a subset $\zeta_s^A$ of the modified contact process $\overline{\xi}_s^A$. To do this, we replace the displacements $U_i$ of particles in the branching random walk, which are uniformly distributed on $[-1, 1]^d$, by $\varphi_M\left(U_i\right)$ where

$$\varphi_M(x) = \pi\left(x\left(M + \tfrac{1}{2}\right)\right)/M$$

and $\pi(x)$ is the closest point in $\mathbb{Z}^d$ to $x$ (with some convention for breaking ties). We will not worry about what to do after a collision (i.e., a birth onto an occupied site) occurs because (9) implies that the probability of this event $\to 0$ as $M \to \infty$. When $V_T \leq M^{1/3}$ and there are no collisions, then all the points of $\zeta_s^A$ end up within $M^{1/3} \cdot 1/M$ ( in $\| \|_\infty$ norm) of their counterparts in the branching process. Since $\overline{Z}_s^A \subset [-2L+1, 2L-1]^d$ for

$0 \le s \le T$, $\zeta_T^A \subset \bar{\xi}_T^A$. $\bar{\xi}_T^A$ may be larger because in that process particles are only killed when they land outside $(-2L, 2L)^d$ but that is not important, so we have shown

(10) Let $\epsilon > 0$ and $K$ the constant chosen in (8). If $M$ is large then for all $A \subset I_0$ with $|A| \in [K, 2K]$, $P\left(\left|\bar{\xi}_T^A \cap I_1\right| < K\right) \le 2\epsilon$.

**C. Block Construction.**

Let $\bar{\xi}_t^{m,n,A}$ be a modified contact process that starts with $A$ occupied at time $nT$ and in which births outside $2mLe_1 + [-2L, 2L]^d$ are not allowed. From (10), translation invariance, and reflection symmetry, it follows that if $A \subset I_m$ has $|A| \in [K, 2K]$ then with high probability $\bar{\xi}_{(n+1)T}^{m,n,A}$ will have at least $K$ points in $I_{m-1}$ and $I_{m+1}$. Since the events associated with different $(m, n) \in \mathcal{L}$ use different parts of the graphical representation it should be clear that (0) holds. To prove this we will use induction. To carry out the proof it is useful to observe that in determining whether or not $|\mathcal{C}| = \infty$ we only need to look at $\omega(m, n)$ for $-n \le m \le n$, so we will only define those variables. Set $\omega(0, 0) = 1$ if $\left|\bar{\xi}_T^{0,0,A} \cap I_i\right| \ge K$ for $i = +1$ and $-1$, and let

$$A_{0,0}^i \subset \bar{\xi}_T^{0,0,A} \cap I_i$$

have cardinality $K$. Otherwise set $\omega(0, 0) = 0$ and $A_{0,0}^i = \phi$ for $i = +1$ and $-1$. In the second case the rest of the construction is rather pointless but we will continue with it so that all the variables $\omega(m, n)$ are always defined.

Assuming the $A_{m,n}^i$ $-n \le m \le n$ have been defined we set

$$A_{-n-1,n}^{+1} = A_{-n+1,n}^{-1} = \phi$$

and

$$A_{m,n+1} = A_{m-1,n}^{+1} \cup A_{m+1,n}^{-1}.$$

There are two cases to consider:

**Case 1.** $A_{m,n+1} \ne \phi$. We set $\omega(m, n+1) = 1$ if

$$\left|\bar{\xi}_{(n+2)T}^{m,n+1,A(m,n+1)} \cap I_{m+i}\right| \ge K \text{ for } i = +1, -1$$

and let

$$A_{m,n+1}^i \subset \bar{\xi}_{(n+2)T}^{m,n+1,A(m,n+1)} \cap I_{m+i}$$

have cardinality $K$. Otherwise set $\omega(m, n+1) = 0$ and $A_{m,n+1}^i = \phi$.

**Case 2.** $A_{m,n+1} = \phi$. We let $H(m) = 2mLe_1 + H$ where $H$ is a fixed subset of $I_0$ of cardinality $K$ and set $\omega(m, n+1) = 0$ if

$$\left|\bar{\xi}^{m,n+1,A(m,n+1)}_{(n+2)T} \cap I_{m+i}\right| \geq K \text{ for } i = +1, -1$$

and $= 0$ otherwise. In either case we set $A^i_{m,n+1} = \phi$ for $i = +1, -1$. To see the reason for these definitions observe that it follows by induction that

$$A_{m,n} \subset \xi^A_{nT} \text{ and } \mathcal{C} = \{(m,n) : A_{m,n} \neq \phi\} \tag{11}$$

If we let $\mathcal{F}_t$ be the $\sigma$-field generated by the graphical representation up to time $t$ then for $(m, n+1) \in \mathcal{L}$ and $n \geq 1$

$$p(\omega(m, n+1) = 1 \mid \mathcal{F}_{nT}) \geq 1 - 4\varepsilon,$$

and given $\mathcal{F}_{nT}$, $\omega(m, n+1), (m, n+1) \in \mathcal{L}$ are independent. From the last result and induction it follows that $\omega(m, n)$ dominates an independent system with density $1 - 4\varepsilon$ and the desired result follows from (2.1).

## 5. Proof of Theorem 4.

We begin by constructing the process from a collection of Poisson processes. For $i \in \{1, 2\}$ and $x, y \in \mathbb{Z}^d/M$ with $\|y - x\|_\infty \leq 1$, let $\{T^{i,x,y}_n : n \geq 1\}$ and $\{U^{i,x}_n : n \geq 1\}$ be the arrival times of Poisson processes with rates $\lambda_i$ and 1 respectively. As the reader can probably guess from the rates: at times $U^{i,x}_n$ we kill the particle at $x$ if it is of type $i$, and at times $T^{i,x,y}_n$ there is a birth from $x$ to $y$ if $x$ is in state $i$ and $y$ is in state $j < i$. Generalizing the usual practice in the graphical representation of the contact process we write a $\delta_i$ at $(x, U^{i,x}_n)$ and draw an arrow of type i from $(x, T^{i,x,y}_n)$ to $(y, T^{i,x,y}_n)$. If, for example $i = 2$, we will call the last object a 2-arrow. To distinguish the two ends of the arrow we will say that it attacks $y$ and its source is $x$.

Even though there are infinitely many Poisson processes, and hence no first arrival, it is easy to modify the argument in Section 3 to show that the recipe above allows us to construct the process starting from any $\xi_0 \in \{0, 1, 2\}^{\mathbb{Z}^d/M}$. We consider a random graph in which $x$ and $y$ are connected if

$$\min\left(T^{1,x,y}_1, T^{1,x,y}_1, T^{1,y,x}_1, T^{2,y,x}_1\right) \leq \tau.$$

If $\tau$ is chosen small enough so that the probability of connection is less than $1/(2M+1)^d$ then comparing with a branching process shows that all the components of our random graph are finite. The evolution of each component is unaffected by the others and can be computed separately. In this way we can construct the process up to time $\tau$ and iterating we can construct the process for all time.

Let $\mathcal{L} = \{(m,m) \in \mathbb{Z}^2 : m+n \text{ is even}\}$, $e_1 = (1,0,...0)$, and

$$\varphi(m,n) = (2mLe_1, nT) \quad \text{where } T = L^2$$
$$B = (-2L, 2L)^d \times [0, L^2] \quad B_{m,n} = \varphi(m,n) + B$$
$$I = [-L, L]^d \quad I_m = 2mLe_1 + I$$

Let $H = \{(1/M)e_1, \dots (N/M)e_1\}$, and $\eta_t^H = \{y : \xi_t^H(y) = 1\}$ where $\xi_t^H$ is the system starting with 1's on $H$ and 2's on $H^c$. We will say that $(m,n) \in \mathcal{L}$ is occupied if $|\eta_{nT} \cap I_m| \geq N$. Our aim is to show:

(0) Let $\varepsilon > 0$. If $K, L$ are chosen appropriately then the process of occupied sites dominates independent oriented site percolation with $p = 1 - \varepsilon$.

Here and throughout the section we are supposing that $\lambda_1 > \lambda_2^2 > 1$ . Once this is established the desired conclusion follows easily from (2.2); see part D of this section. To prove (0) we will prove

(*) Let $\varepsilon > 0$. If $K, L$ are chosen appropriately and we have $A \subset I_m$ with $|A| \geq N$ occupied by 1's then, even if there are 2's at all the sites in $A^c$ at time $nT$ and at all sites in $B_{m,n}^c$ at all times $t \in [nT, (n+1)T]$, we will with probability $\geq 1 - \varepsilon$ have at least $N$ sites occupied by 1's in $I_{m+1}$ and $I_{m-1}$.

We assume "there are 2's at all the sites in $A^c$ at time $nT$ and at all sites in $B_{m,n}^c$ at all times $t \in [nT, (n+1)T]$" to make what happens in the different $B_{m,n}$ independent. Before embarking on the proof of (0) we would like to observe that by translation invariance it suffices to prove the result when $m = 0$ and $n = 0$.

**A. Preliminaries.**

To get the real system from the mean field system defined in the introduction, we will have to replace the coin flips by "*2-dual processes*", so we begin by describing that notion. Suppose $M < \infty$ and let $\hat{\zeta}_s^{(x,t)}$ = the set of points at time $t - s$ that can be reached by a *"dual 2 path"* starting from $(x, t)$. These paths can go down the graphical representation (but not through $\delta_2$'s) and across 2-arrows in the direction opposite to their orientation. It is easy to check (see Durrett (1988) Chapter 4) that

$$\{x \in \zeta_t\} = \left\{\hat{\zeta}_t^{(x,t)} \cap \zeta_0 \neq \phi\right\}$$

$$\left\{\hat{\zeta}_s^{(x,t)} : 0 \le s \le t\right\} \stackrel{d}{=} \{\zeta_s^x : 0 \le s \le t\}.$$

We will decide that $x$ is occupied by a 2 at time $t$ if $\hat{\zeta}_s^{(x,t)}$ survives until time $K$ or reaches $(x + [-K, K]^d)^c$. We have to truncate in time and space to get independence of what happens in different blocks. The first step is to determine how large $K$ should be. In subsection $C$ we will show

(1) As $M \to \infty, \hat{\zeta}_s^{(x,t)}$ approaches a branching random walk $Y_s^x$ in which (i) Particles die at rate 1 and give birth at rate $\lambda_2$. (ii) The offspring of a particle at $x$ is sent to a point $y$ chosen at random from $x + [-1, 1]^d$.

Pick $\rho > (\lambda_2 - 1)$ so that $1 + \lambda_2 \cdot \rho < \lambda_1/\lambda_2$. The survival probability $P(Y_s^x > 0 \text{ for all } s \ge 0) = (\lambda_2 - 1)/\lambda_2$ so

(2) If we pick $K$ large then the probability $Y_s^x$ survives until time $K$ or reaches a point in $(x + [-K+1, K-1]^d)^c$ is $< \rho$.

The +1 and -1 are to leave room for the limit $M \to \infty$. The last observation will imply that after time $K$, or for sites in $[-2L+K, 2L-K]^d$, the collection of 2's that we see is, for large $M$, not too much thicker than a product measure with density $\rho$. We can avoid the sites near the boundary of $[-2L, 2L]^d$ by not using them. We take a rather drastic approach to cope with the first $K$ units of time. We ignore births and observe that each 1 dies at rate at most $\lambda_2 + 1$ so

(3) Let $\epsilon > 0$. If $N \ge N_0$ then with probability $\ge 1 - \epsilon$ the number of surviving 1's at time $K$ is least $N' \equiv [\frac{N}{2} \cdot \exp(-(\lambda_2 + 1)K)] + 1$.

Here $[x]$ = the largest integer $\le x$.

The first phase of the construction decimates the set of 1's, but after that phase the set of 1's will almost be a supercritical branching random walk, so we can recover our losses. We begin by considering what happens when $M = \infty$. Let $\bar{Z}_t^x$ be a modification of the mean field process starting from a single 1 at $x$, in which particles die at rate $1 + \lambda_2\rho$ and those that land outside $[-2L+K+1, 2L-K-1]^d$ are killed. It follows from arguments in Section 4 (see (1)–(7)) that

4) Let $\varepsilon > 0$ and $T' = L^2 - K$. If $L$ is large,

$$\inf_{x\in[-L,L]^d} E\left|\bar{Z}^x_{T'} \cap I'\right| \geq 4\exp\left((\lambda_2 + 1)K\right).$$

If $N \geq N_1 (\geq N_0)$ then for all $A' \subset [-L, L]^d$ with $|A'| \geq N'$ (the quantity in (3))

$$P\left(\left|\bar{Z}^{A'}_{T'} \cap I'\right| < N\right) \leq \varepsilon.$$

**B. Block Construction.**

Given a subset $A$ of $I_m$, we will define a process $\bar{\eta}^{m,n,A}_t$ that is a subset of the sites occupied by 1's at time $t$ when we start with A occupied by 1's at time nT. $\bar{\eta}^{m,n,A}_{nT} = A$ and evolves as follows:

(i) For $nT \leq t \leq nT + K$ if a $\delta_1$ lands at $x$, or $x$ is attacked by a 2-arrow, it is removed from the set. In this phase births of 1's are ignored.

(ii) For $nT + K \leq t \leq (n+1)T$, if a $\delta_1$ lands at $x \in \bar{\eta}^{m,n,A}_t$ it is removed from the set. If there is a 1-arrow from $x \in \bar{\eta}^{m,n,A}_t$ to $y$, we look at $\hat{\zeta}^{(y,t)}_s$. If the dual process survives until time $K$ or reaches $\left(y + [-K, K]^d\right)^c$ we decide that the site is occupied by a 2 and $y$ is not added. Otherwise it is added.

If there is a 2-arrow from $y$ to $x \in \bar{\eta}^{m,n,A}_t$ then we look at $\hat{\zeta}^{(y,t)}_s$. If the dual process survives until time $K$ or reaches $\left(y + [-K, K]^d\right)^c$ we decide that the site is occupied by a 2 and we remove $x$ from $\bar{\eta}^{m,n,A}_t$. Otherwise nothing happens.

Let $A' = \bar{\eta}^{m,n,A}_{nT+K}$. (3) tells us that if $|A| \geq N$ then with high probability $|A'| \geq N'$. In subsection $C$ we will show

(5) As $M \to \infty$ the behavior of $\bar{\eta}^{m,n,A}_{nT+K+t}$ approaches that of $\bar{Z}^A_t$, so $(*)$ holds.

As we will now explain, this allows us to prove (0). The argument is similar to the one in part $C$ of the last section. To carry out the proof, it is useful to observe that in determining whether or not $|\mathcal{C}| = \infty$ (see Section 2 for the definitions of this and other quantities associated with the percolation process) we only need to look at $\omega(m, n)$ for $(m, n) \in \mathcal{L}$ with $-n \leq m \leq n$, so we will only define those variables.

Let $A_{0,0} = H = \{(1/M)e_1, \ldots (N/M)e_1\}$. Assuming the $A_{m,n}$, $-n \leq m \leq n$, have been defined, we will now define the $\omega(m, n)$. There are two cases to consider:

**Case 1.** $A_{m,n} \neq \phi$. We set $\omega(m,n) = 1$ if $\left|\bar\eta_{(n+1)T}^{m,n,A(m,n)} \cap I_{m+i}\right| \geq N$ for $i = +1, -1$ and let

$$A_{m,n}^{i} \subset \bar\eta_{(n+1)T}^{m,n,A(m,n)} \cap I_{m+i}$$

have cardinality $N$. Otherwise set $\omega(m,n) = 0$ and $A_{m,n}^{i} = \phi$.

**Case 2.** $A_{m,n} = \phi$. We let $H(m) = 2mLe_1 + H$ and set $\omega(m,n) = 1$ if

$$\left|\bar\eta_{(n+1)T}^{m,n,H(m)} \cap I_{m+i}\right| \geq N \text{ for } i = +1, -1$$

and $\omega(m,n) = 0$ otherwise. In either case we set $A_{m,n}^{i} = \phi$ for $i = +1, -1$.

To make the next generation of A's we set

$$A_{-n-1,n}^{+1} = A_{n+1,n}^{-1} = \phi \text{ and } A_{m,n+1} = A_{m-1,n}^{+1} \cup A_{m+1,n}^{-1}.$$

To see the reason for the last set of definitions observe that it follows by induction that

$$(6) \qquad A_{m,n} \subset \eta_{nT}^{H} \text{ and } C = \{(m,n) : A_{m,n} \neq \phi\}.$$

From the definitions of the boxes, it follows that if we let $\mathcal{F}_t$ be the $\sigma$-field generated by the graphical representation up to time $t$ then for $(m,n) \in \mathcal{L}$

$$P(\omega(m,n) = 1 \,|F_{nT}) \geq 1 - \varepsilon,$$

and given $\mathcal{F}_{nT}$, $\omega(m,n)\, m \in \{-n, -n+2, \ldots n\}$ are independent.

**C. Continuity argument.**

In this section we will prove (1) and (5). We begin with the proof of (1) which is almost the same as the argument in part B of Section 4. The number of points in

$$V_r \equiv \bigcup_{0 \leq s \leq r} \hat\zeta_s^{(x,t)}$$

is dominated by a branching process $W_r$ in which births occur at rate $\lambda_2$ and deaths at rate 0. $EW_t = \exp(\lambda_2 t) < \infty$, so $P\left(W_K > M^{1/3}\right) \to 0$. If there are $\leq M^{1/3}$ points in $V_K$ then the probability of having a birth land on an occupied site is smaller than

$$(7) \qquad M^{1/3}\left(M^{1/3}/(2M+1)^d\right) \to 0.$$

To deal with the spatial location of the particles, we observe that given a realization of the branching random walk $Y_s^x, s \geq 0$, defined in (1), we can construct a realization of the 2-dual by replacing the displacements $U_i$, which are uniform on $[-1,1]^d$, by $\pi\left(\left(M+\frac{1}{2}\right)U_i\right)/M$ where $\pi(y)$ is the closest point in $\mathbb{Z}^d$ to $x$ (with some convention for breaking ties). We will not worry about what to do after a collision (a birth onto an occupied site) occurs because (7) implies that the probability of a collision before time $K$ approaches 0 when $M \to \infty$. When there are $\leq M^{1/3}$ points in $V_K$ and no collision occurs, then all the particles in $\hat{\zeta}_K^{(x,t)}$ end up within $M^{1/3}/M$ (in $L^\infty$ norm) of their counterparts in $Y_K^x$. This proves (1) and in combination with (2) proves

(8) If $M$ is large the probability that $\hat{\zeta}_s^{(x,t)}$ survives until time $K$ or reaches $(x+[-K,K]^d)^c$ is $< \rho$.

The last argument contains all the ideas for showing that the behavior of $\eta_{nT+K+t}^{m,n,A'}$ approaches that of $Z_t^{A'}$ as $M \in \infty$. In the event of interest we start with $\leq 2N$ particles, so comparing with a branching process $W_t$ in which particles are born at rate $\lambda_1$ and die at rate 0, we conclude that the probability of a newborn 1 landing on a site in $\eta_{nT+K+t}^{m,n,A}$ approaches 0 as $M \to \infty$. When $W_{T'} \leq M^{1/5}$ new 2-duals, which we have to follow backwards, are generated at rate $\leq \lambda_2 M^{1/5}$ so with high probability, $\leq 2\lambda_2 M^{1/5}$ duals are generated. By computing second moments and using Chebyshev's inequality we see that if M is large then with high probability

$$\left|\bigcup_{s=0}^{T} \hat{\zeta}_s^{(x,t)}\right| \leq M^{1/5}$$

for all these duals. The total number of sites we ever look at is $\leq CM^{2/5}$ so the probability of a collision is smaller than

$$CM^{2/5}\left(CM^{2/5}/\left(2M+1\right)^d\right) \to 0$$

By considering the locations of the particles involved as before, one concludes that (5) holds and (0) follows.

**D. Denouement.**

The last detail is to produce the desired stationary distribution. The first step is to take Cesaro averages of the distributions of $\eta_s^H$, $0 \le s \le 2nT$, where $H = \{(1/M)\, e_1, \dots (N/M)\, e_1\}$, let $n \to \infty$, and extract a convergent subsequence. This is possible since the set of probability measures on $\{0,1,2\}^{\mathbb{Z}^d}$ is compact (in the obvious topology). Let $\mu$ be a subsequential limit. Since our process has the Feller property, $\mu$ is a stationary distribution. (See Liggett (1985), p. 10). Let $A_0, A_f, A_\infty$ be the set of configurations in $\{0,1,2\}^{\mathbb{Z}^d}$ with 0, finitely many, and $\infty$ many ones respectively. Since $\mu$ is a stationary distribution $\mu(A_f) = 0$, and so it follows from (2.2) that $\mu(A_\infty) > 0$. Let $\mu_{12}(B) = \mu(B \cap A_\infty) / \mu(A_\infty)$. Since $A_0$ and $A_\infty$ are invariant for the process, $\mu_{12}$ is a stationary distribution. By definition $\mu_{12}(A_\infty) = 1$. To see that $\mu_{12}$ concentrates on configurations with infinitely many 2's we observe that an easy argument shows that $\zeta_t = \{y : \xi_t^H(y) = 2\} \Rightarrow \mu_2$. (Use the self-duality of the contact process and observe that the probability the number of particles in the dual is in $[1, N]$ goes to 0 since $\phi$ is an absorbing state.)

## 6. Proof of Theorem 2

We begin by constructing the process from a collection of Poisson processes. For $x, y \in \mathbb{Z}^d$ with $|x - y| = 1$ let $\{T_n^{x,y} : n \ge 1\}$ and $\{U_n^x : n \ge 1\}$ be the arrival times of Poisson processes with rates 1 and $\delta$ respectively. As the reader can probably guess from looking at the rates: at times $U_n^x$ we kill the particle at $x$ if one is present, and at times $T_n^{x,y}$ there is a birth from $x$ to $y$ if $x$ is occupied. For reasons indicated in Section 3, the above recipe allows us to construct the process starting from any initial $\xi_0$.

Let $\varphi(m,n) = (2mLe_1, nT)$, where $T = \kappa_d L$, and $\kappa_d$ is a constant that depends on the dimension and will be chosen later. Let

$$B = (2L, 2L)^d \times [0, T] \qquad B_{m,n} = \varphi(m,n) + B$$
$$I = [-L, L]^d \qquad I_m = 2mLe_1 + I$$

We will say that $(m,n) \in \mathcal{L}$ is occupied if $\xi_{nT} \cap I_m \ne \phi$. Our aim is to show:

(0) Let $\varepsilon > 0$. If L is large the process of occupied sites dominates independent oriented site percolation with $p = 1 - \varepsilon$.

As in Section 3, it suffices to prove this when $\delta = 0$, for once $L$ and $T$ are fixed, it follows that when $\delta \le \delta_0(\varepsilon)$ the probability of a death in $B_{m,n}$ is $\le \varepsilon$ and (0) holds with $\varepsilon$ replaced by $2\varepsilon$.

For the rest of the section we will restrict our attention to the case $\delta = 0$. The key to the proof of (0) in this case is defining a tagged particle $r_t \in \xi_t$ that, together with a mirror image (the reflection in the hyperplane perpendicular to the first axis), will be used to move a particle from $I_m$ to $I_{m+1}$ and one from $I_m$ to $I_{m-1}$. In what follows, it is convenient to use function notation for the process, i.e., $\xi_t(x) = 1$ if $x \in \xi_t$ and $= 0$ otherwise.

To warm up for the real definition we begin by considering $d = 1$. In this case things will be arranged so that at all times $\xi_t(r_t) = 1$ and $\xi_t(r_{t+1}) = 0$. We do not move the particle until the first time that $\xi_t(r_t) = 0$ or $\xi_t(r_t{+}1) = 1$. In the first case $\xi_t(r_t - 1) = 1$ so we move to $r_t - 1$. In the second case we move to $r_t + k$ where $k \geq 1$ is the smallest positive integer for which $\xi_t(r_t + k) = 1$ and $\xi_t(r_t + k + 1) = 0$. The tagged particle moves forward by $k \geq 1$ at rate 1, moves backwards by 1 at rate 1 when $\xi_t(r_t - 1) = 1$, and moves backwards at rate 0 when $\xi_t(r_t - 1) = 0$. Therefore if we can show $\xi_t(r_t - 1) = 0$ a positive fraction of the time, $r_t$ will move to the right at a positive rate.

In defining $r_t$ in $d > 1$, we want the first coordinate to increase linearly and keep the other coordinates close to 0. Things will be arranged so that at all times

(C0): $\xi_t(r_t) = 1$

(C1): $\xi_t(r_t + e_1) = 0$.

(Ci) $2 \leq i \leq d : \xi_t(r_t - e_i) = 0$ if $r_t^i > 0$, $\xi_t(r_t + e_i) = 0$ if $r_t^i < 0$.

Here $e_i$ is the ith unit vector and $r_t^i$ is the ith coordinate of $r_t$.

We do not move our particle until one of the conditions becomes violated. If it is one of the conditions (C1)-(Cd) that fails we will use the

**A. Repositioning Algorithm.** Repeatedly apply the following rules until $(C1) - (Cd)$ hold:

$(R1)$ If $\xi_t(r_t + e_1) = 1$ then move to $r_t + e_1$.

$(Ri)$ $2 \leq i \leq d$: Let $\alpha_i(x) = x - e_i$ if $x^i > 0$ and $x + e_i$ if $x^i < 0$.

If $\xi_t(\alpha_i(r_t)) = 1$ then we move to $\alpha_i(r_t)$.

If several rules can be applied, use the one with the smallest number.

For the discussion below it is useful to note that if we define $\alpha_1(x) = x + e_1$ then the first rule is a special case of the second.

When $(C0)$ fails, the particle at $r_t$ was killed by a particle

$$y \in \bigcup_{i=1}^{d} A_i(x)$$

where $A_1(x) = \{x - e_1\}$ and for $2 \le i \le d$

$$A_i(x) = \begin{cases} \{x + e_i\} & \text{if } x^i > 0 \\ \{x - e_i\} & \text{if } x^i < 0 \\ \{x + e_i, x - e_i\} & \text{if } x^i = 0. \end{cases}$$

If this happens we move our tagged particle to $y$ and apply the repositioning algorithm. For practice with the rules try the following example:

$$\begin{matrix} 0 & 1 & c & 0 & 0 & 0 \\ 1 & * & 0 & 1 & 1 & 0 \\ 1 & 0 & 0 & 0 & a & 0 \\ 1 & 1 & b & 0 & 0 & 1 \\ 0 & 1 & 0 & 0 & 1 & 0 \end{matrix}$$

Here $*$ indicates the position of the tagged particle, which for concreteness we imagine to be at (2,4). The sites a,b, and c are occupied by 1's and are the possible new locations of the tagged particle.

| event | new position |
|---|---|
| birth occurs at $r_t + e_1$ | a |
| birth occurs at $r_t - e_2$ | b |
| $r_t$ killed by $r_t - e_1$ | b |
| $r_t$ killed by $r_t + e_2$ | c |

Intuitively $r_t^1$ has positive drift for the following reasons:

(i) When $\xi_t(r_t - e_1) = 1, r_t^1 \to r_t^1 - 1$ at rate 1 and $r_t^1 \to r_t^1 + k$ with $k \ge 1$ at rate $\ge 1$.
(ii) When $\xi_t(r_t - e_1) = 0, r_t^1 \to r_t^1 - 1$ at rate 0 and $r_t^1 \to r_t^1 + k$ with $k \ge 1$ at rate $\ge 1$.
(iii) We move from case (i) to case (ii) at rate $\ge 1$ (i.e., when the particle at $r_t$ kills the one at $r_t - e_1$) and from case (ii) to case (i) at rate $\le 2d(1+d) + d - 1$ (i.e., $r_t - e_1$ gets filled in at rate $\le 2d$ but if the tagged particle moves because one of the conditions $(C0) - (Cd)$ becomes violated we may end up in case (i) . Each of the $d$ points $\alpha_i(r_t)$ could get filled in at rate 2d. The particle at $r_t$ gets killed at rate $\le d - 1$.)

To translate the intuition just explained into a proof, we will define point processes on $\{+\} \times [0, \infty)$ and $\{-\} \times [0, \infty)$. When there is a birth from $r_t$ to

$r_t + e_1$ we will put a point at $(+, t)$. Let

$$\varphi(t) = \int_0^t 1_{(\xi_s(r_s - e_1) = 1)} ds$$

and when the particle at $r_t$ is killed by one at $r_t - e_1$ put a point at $(-, \varphi(t))$. It is easy to see that the processes just defined are rate one Poisson processes. The $+$ and $-$ are to indicate that at the corresponding times $r_t^1$ changes by $\geq +1$ and $\geq -1$ respectively, so if we let $N_s^+$ and $N_s^-$ be the number of points in $[0, s]$ in the two processes then

$$r_t^1 \geq S_t \equiv N_t^+ - N_{\varphi(t)}^-.$$

Our first step in getting a lower bound on $S_t$ is to get an upper bound on $\varphi(t)$. From (iii), we see that $\xi_t(r_t - e_1)$ stays 1 for an amount of time $\leq$ an exponential with mean 1 and stays 0 for an amount of time $\geq$ an exponential with mean $1/\{2d(d+2) - 1\}$. A simple argument shows

$$\limsup_{t\to\infty} \varphi(t)/t \leq 1 - (1/2d(d+2)),$$

and furthermore

(1) If $a > 1 - (1/2d(d+2))$ there are constants $C, \gamma$ that depend on $a$ so that

$$P(\varphi(t)/t > a) \leq Ce^{-\gamma t}.$$

Combining the last result with large deviations results for the Poisson process gives

(2) If $b < 1/(2d(d+2))$ there are constants $C, \gamma$ that depend on b so that

$$P(S_t < bt) \leq Ce^{-\gamma t}.$$

Using the fact that $S_t \geq -N_{L/2}^-$ for all $t \geq L/2$, and summing the last estimate with $b = 1/4d(d+2)$ over integers $t \geq L/2$ leads easily to

(3)

$$P\left(\inf_t S_t < -L\right) \leq Ce^{-\gamma L}.$$

Here and in what follows $C, \gamma \in (0, \infty)$ but change from line to line.

The above arguments can also be used to bound the size of $r_t^2, ... r_t^d$. To do this we observe that the same intuition implies that when $r_t^j \leq 0$, it has positive drift. To turn our intuition into a proof we define $N_t^+, \varphi(t), N_t^-$ as before but with 1 replaced by $j$.

**B. Construction.**

The goal of our construction is to "move" a particle from $I_m$ to $I_{m+1}$ by time $(n+1)T$. To carry out the proof, it is convenient (and sufficient for the proof) to consider the case $m = 0, n = 0$. Before entering into the somewhat unpleasant details, we would like to point out the source of our trouble: The arguments in the last section gives us a lower bound on the drift to the right but no upper bound. In earlier arguments, e.g., in Section 3, we could use the strong law of large numbers to force the particle to be in the right place at the right time. Here the tagged particle may move at different speeds when the initial state $A = \{0\}$ or $A = \mathbb{Z}^d$, so our construction has to be more complicated.

The first step in the construction is to find a particle $\rho_0$ so that $\alpha_i(\rho_0)$ is vacant for $1 \le i \le d$. Let $x_0 \in A$ and define a sequence of points as follows. If all the $\alpha_i(x_k)$ are vacant we stop. If at least one of the points $\alpha_i(x_k)$ is occupied let $x_{k+1}$ be the one with the smallest value of $i$ and repeat. One of two things can happen: (i) the construction terminates at a point with the desired properties without leaving $[-(3/2)L, (3/2)L]^d$ or (ii) not. In the second case let $y_k$ be the first $x_i$ not in $(-L-4k, L+4k)^d$, let $\eta_k = \{\alpha_i(y_k) : 1 \le i \le d\} \cap A$ and let $G_k = \{$ from time 0 to 1 there is exactly one arrow from $y_k$ to each point in $\eta_k$ and no other arrows land on $\{y_k\} \cup \eta_k\}$. The events $G_k$, $1 \le k \le [L/8]$ are independent and each has probability $\exp(-2d(d+1))$, so with probability at least

$$(4) \qquad \{1 - \exp(-2d(d+1))\}^{[L/8]}$$

one of these good events will occur and give us a place to start our construction.

In case (i) our particle $\rho_t$ starts moving at time 0; in case (b) at time 1. In either case our particle $\rho_t$ starts at the location found in the last paragraph and behaves like the tagged particle $r_t$ until time

$$\tau_1 = \inf\left\{t : \rho_t^1 \ge 3L/2\right\}.$$

To keep the particle from flying out of the box $B_{0,0}$ at time $\tau_1$, we stop the repositioning step at time $\tau_1$ when the first coordinate becomes $3L/2$. (We assume $L$ is even.)

If $\tau_1 \le T$ then at time $\tau_1$ we have achieved our goal of moving the first coordinate into $[L, 2L]$ and the construction enters its second phase. Let

$\beta_1(x) = x - e_1$ if $x^1 > 3L/2$, $\beta_1(x) = x - e_1$ if $x < 3L/2$. If $2 \le i \le d$ let $\beta_1(x) = \alpha_i(x)$. During this part of the construction, things will be arranged so that at all times.

$$(\bar{C}0) : \eta_t(r_t) = 1$$
$$(\bar{C}i)\ 1 \le i \le d : \eta_t(\beta_i(r_t)) = 0.$$

We do not move our particle until one of the conditions becomes violated. If it is one of the conditions $(\bar{C}1) - (\bar{C}d)$ that fails we will use.

**C. Repositioning Algorithm II.** Repeatedly apply the following rules until $(\bar{C}1) - (\bar{C}d)$ hold:

$$(\bar{R}i)\ 1 \le i \le d: \quad \text{If } \eta_t(\beta_i(r_t)) = 1 \text{ then we move to } \beta_i(r_t).$$

If several rules can be applied, use the one with the smallest number.

Since each move brings the particle closer to $(3L/2, 0, ...0)$ the algorithm stops after a finite number of steps. Notice that now the first coordinate is treated like the others (except for the fact that we try to keep it near $3L/2$).

To prove the construction suceeds, we begin by observing that the first coordinate of our particle starts at $\rho_0 > -3L/2$, and our goal is to get to $3L/2$, so if $\kappa_d = 2 \cdot 3 \cdot 2d(d+2)$ then (3.2) and (3.3) imply that with high probability (i.e., with a probability $\to 1$ as $L \to \infty$)

$$\tau_1 < T = \kappa_d L \text{ and } \rho_t^1 > -2L \text{ for } t \le \tau_1.$$

Turning now to the behavior of $\rho_t^i, 2 \le i \le d$, we begin by observing that $\rho_0^i$ is in $[-3L/2, 3L/2]$, so the arguments for $\rho_t^1, 0 \le t \le \tau_1$ show that if $\tau_i^0 = \inf\{t : \rho_t^i = 0\}$ then with high probability

$$\tau_i^0 < T \text{ and } |\rho_S| \le 2L \text{ for } s \le \tau_i^0.$$

For $2 \le i \le d$ let

$$\sigma_i^k = \inf\left\{t > \tau_i^{k-1} : \rho_t^i \ne 0\right\}$$
$$\tau_i^k = \inf\left\{t > \sigma_i^{k-1} : \rho_t^i = 0\right\}$$

and $K = \inf\{k : \sigma_i^k > T\}$. Every time $\rho_t^i = 0$, it stays there for at least an exponential amount of time with mean $1/2$ (i.e., until $\rho_t$ is killed by a particle at $\rho_t + e_i$ or $\rho_t - e_i$), so with high probability $K \leq 4T = 4\kappa_d L$. An easy generalization of (3) shows

$$P\left(\left|\rho_t^i\right| \geq L \text{ for some } t \in \left[\sigma_i^k, \tau_i^k\right]\right) \leq Ce^{-\gamma L}.$$

Combining the last two results we conclude that with high probability, $\left|\rho_t^i\right| \in (-L, L)$ for all $t \in [\tau_i^0, T]$. For the case $i = 1$, replacing 0 by $3L/2$ in the definition of $\sigma_i^k$ and $\tau_i^k$ for $i = 1$ and setting $\tau_i^0 = \tau_1$, the last argument also shows that with high probability $\rho_t^1 \in (L, 2L)$ for $t \in [\tau_1, T]$. This completes the proof that our procedure moves a particle from $I_0$ to $I_1$ with high probability. This leads easily to (0) and completes the proof of Theorem 2.

## References

Theorem 1A was first proved in [11]. The proof given here is based on the one in [2]. Theorem 1B is stated for ease of exposition only. It is not the best known result on the contact process so its proof is left as an exercise for the reader. See Section 5 of [8] for help. Theorems 2A and 2B are from [2]. Theorem 3A is from [4]. Proofs of Theorem 3A and 3B can be found in [16]. Theorem 4A is from [8]. The proof of Theorem 4B is the proposed topic of Anne Olsson's Ph.D. thesis. Since this paper was written, Bezuidenhout and Grimmett have shown that the complete convergence theorem (i.e., the conclusion of theorems 1B and 3B) holds for the contact process for any range and in any $d \geq 1$.


[1] C. Bezuidenhout and G. Grimmett, *The critical contact process dies out*, Ann. Probab. **18** (1990), pp. 1462–1482.

[2] M. Bramson, W.D. Ding and R. Durrett, *Annihilating branching processes*, Preprint.

[3] M. Bramson and R. Durrett, *A simple proof of the stability theorem of Gray and Griffeath.*, Prob. Th. Rel. Fields **80** (1989), pp. 293–289.

[4] M. Bramson, R. Durrett, and G. Swindle, *Statistical mechanics of crabgrass*, Ann. Probab. **17** (1989), pp. 444–481.

[5] M. Bramson and L. Gray, *The survival of branching annihilating random walk*, Z. fur Wahr **68** (1985), pp. 447–460.

[6] R. Durrett, *Oriented percolation in two dimensions*, Ann. Probab. **12** (1984), pp. 999–1040.

[7] R. Durrett, "Lecture Notes on Particle Systems and Percolation," Wadsworth, Pacific Grove, CA, 1988.

[8] R. Durrett and R.H. Schonmann, *Stochastic growth models*, Percolation and the Ergodic Theory of Interacting Particle Systems, in "IMA volumes in Math. and its Appl.," Springer Verlag, New York, 1987.

[9] R. Durrett and G. Swindle, *Are there bushes in a forest?*, Stoch. Proc. Appl. **37** (1991), pp. 19–31.

[10] L. Gray and D. Griffeath, *A stability criterion for attractive nearest-neighbor spin systems on* $\mathbb{Z}$, Ann. Probab. **10** (1982), pp. 67–85.

[11] T.E. Harris, *Nearest-neighbor Markov interaction processes on multidimensional lattices*, Adv. Math. **9** (1972), pp. 66–89.

[12] T.E. Harris, *Contact interactions on a lattice*, Ann. Probab. **2** (1974), pp. 969–988.

[13] T.E. Harris, *On a class of set-valued Markov processes*, Ann. Probab. **4** (1976), pp. 175–194.

[14] T.E. Harris, *Additive set-valued Markov processes and graphical methods*, Ann. Probab. **6** (1978), pp. 355–378.

[15] T.M. Liggett, "Interacting Particle Systems," Springer, New York, 1985.

[16] A.M. Olsson, *Master's thesis*, Aarhus University.

Partially supported by the National Science Foundation, the Army Research Office through the Mathematical Sciences Institute at Cornell, and a Guggenheim fellowship. The first draft of this paper was based on a series of seven two-hour lectures given in Sept. 1988 at the Nankai Institute for Mathematics, Tianjin, China.

Department of Mathematics
Cornell University
Ithaca, New York 14853

# Cyclic Cellular Automata in Two Dimensions

Robert Fisch, Janko Gravner and David Griffeath

## Abstract

Start by randomly populating each site of the two-dimensional integer lattice with any one of $N$ types, labeled $0, 1, \ldots, N-1$ ($N \geq 3$). The type $\zeta(y)$ at site $y$ can *eat* the type $\zeta(x)$ at neighboring site $x$ (i.e., replace the type at $x$ with $\zeta(y)$) provided that $\zeta(y) - \zeta(x) = 1 \bmod N$. We describe the dynamics of *cyclic cellular automata* (c.c.a.) $\zeta_t$, discrete-time deterministic systems which follow the rule:

($\bullet$) At any time $t$, each type $\zeta_t(y)$ eats every neighboring type that it can.

These systems have remarkably complex dynamics. As $N$ becomes large they display a curious metastability leading to large-scale locally-periodic structure. This article contains a preliminary account of our findings. For the most part, we rely on computations and computer graphics produced by the Cellular Automaton Machine. However we are able to give a simple proof that the infinite system $\zeta_t$ is asymptotically locally periodic for any $N < \infty$. Moreover, we identify a number of regularity properties of rule ($\bullet$), mostly topological in nature, that offer some hope for a more detailed rigorous analysis.

## 1. Introduction

Think of the two-dimensional integers $\mathbb{Z}^2$ as a terrain inhabited by creatures of $N$ *types*, labeled $0, 1, \ldots, N-1$ ($N \geq 3$). Let $\zeta_0(x)$, $x \in \mathbb{Z}^2$, denote an i.i.d. random initial configuration of types, the value at each site $x$ being uniformly distributed over $[0, \ldots, N-1]$. The creatures proceed to compete for space by feeding on their neighbors, but each type can only eat the one immediately preceding it in a "cyclic food chain." More precisely, if $\|y - x\| = 1$, the type $\zeta(y)$ at site $y$ can *eat* the type $\zeta(x)$—that is, replace the type at $x$ with its own—provided that $\zeta(y) - \zeta(x) = 1 \bmod N$. This article describes the dynamics of discrete-time deterministic systems with local transitions of this sort. Namely, we study *cyclic cellular automata* (c.c.a.) determined recursively by the rule:

$$(\bullet) \qquad \begin{aligned} \zeta_{t+1}(x) &= \zeta_t(y) && \text{if } \zeta_1(y) - \zeta_t(x) = 1 \bmod N \text{ for } \textit{some } y \\ & && \text{such that } \|y - x\| = 1, \\ &= \zeta_t(x) && \text{otherwise.} \end{aligned}$$

In other words, at any time $t$, each type eats *every* neighbor that it can. Note that this update rule makes sense since each type can only be eaten by one other.

The c.c.a. $\zeta_t$ determined by ($\bullet$) exhibit unusual self-organization over time, leading to large-scale locally-periodic structure. We will describe four distinct phases of the evolution, characterized by ***debris, droplets, defects*** and ***demons.*** Each of these objects will be defined rigorously, and a series of propositions, essentially topological, will illuminate their roles in the dynamics. The c.c.a. are interesting not only as mathematically tractable prototypes for local periodicity, but also for the exotic *metastability* they display along the way to the limit.

Our goal here is to identify the basic qualitative behavior of these systems, give some background that motivates our work, present a few empirical findings, and lay the groundwork for a rigorous mathematical analysis. A more complete and systematic study will be forthcoming [7]. A popular account of c.c.a. dynamics appeared in [4].

## 2. Four stages of complex dynamics

The c.c.a. $\zeta_t$ form a one-parameter family of random systems indexed by the number of types $N$. Noise in the i.i.d. initial configuration $\zeta_0$ is subjected to the organizational logic of deterministic rule ($\bullet$). As we shall see, this dynamic gives rise to coherent limiting structure which grows in size as $N$ increases. Approximation of the infinite c.c.a. by finite lattice systems is therefore delicate. To obtain an accurate glimpse of a finite window, say of width $L$, in the infinite c.c.a., we must select finite $L$-systems with $L$ a suitably chosen function of $N$. In particular, $L$ must grow very rapidly with $N$.

In order to describe the essential qualitative features of c.c.a. evolution, let us first consider the 14-color model on a $256 \times 256$ finite sublattice of $\mathbb{Z}^2$, with wrap-around at the boundary. The choice $N = 14$ is large enough that all stages of the complex dynamics are plainly visible, but small enough that the finite system with $L = 256$ provides a faithful representation of the infinite 14-type system. Computer graphics, with types represented as *colors*, effectively communicate the behavior of $\zeta_t$. Most of our computations have been performed on a customized Cellular Automaton Machine (CAM) [12]. We use the word *computation* rather than *simulation* because pseudo-randomness only generates the initial state $\zeta_0$. The scenario described below is remarkably insensitive to the statistics of $\zeta_0$, at least in its qualitative aspects.

Sample trajectories of $\zeta_t$, such as the one illustrated in Plates 1–4, are characterized by *four* distinct stages:

**Stage 1** The vast majority of creatures have nothing to eat initially or quickly run out of food—i.e., within a short time there are only a very few types that have a neighbor they can eat. However, the rare remaining active areas form *critical droplets.*

**Stage 2** These *critical droplets* remain active and, by generating waves, expand at a linear rate until they overrun all of the inactive *debris* that was present in the initial state.

**Stage 3** *Defects* are formed, leading to the emergence of periodic spiral structures. Spirals overtake the Stage 2 wave activity at a linear rate, competing with one another for all of the available space.

**Stage 4** Certain minimal defects, which we call *clocks*, give rise to optimally efficient demons. These *demons* displace less efficient spirals until every site in the lattice is regulated by a local, periodic structure of period $N$.

For $N = 14$, Stage 1 is over quite quickly; the droplets form and coalesce so rapidly that it is hard to distinguish one from another. (By merely increasing $N$ to 20, as illustrated in Plate 4, the active droplets become quite distinct and "drop-like.") Plate 1 depicts the situation well into Stage 2: the lattice is about equally divided between fine-grained regions that consist mainly of debris from $\zeta_0$, and waves of solid type.

Plate 2 occurs early in Stage 3. A *first spiral* has "popped up" in the center of the picture. Others appeared shortly thereafter, and still others are in the process of forming. We will explain later how these spiral formations are generated by topological defects.

The state of $\zeta_t$ toward the end of Stage 3 is shown in Plate 3: all waves of solid color have been overwhelmed by spiral formations. Note that the first spiral is still evident in the center of the plate. But note also the spiral with center just to the north and slightly west. As chance would have it, this is a demon—a structure driven by a period 14 clock. The first spiral, on the other hand, turned out to have period 15. The eventual periodic state is reached after several thousand more updates. In it, the first spiral is almost entirely displaced by the nearby demon. Moreover, every site of the entire lattice becomes part of some demon, cycling through the 14 states in order. Different regions are regulated by different clocks that are out of phase with one another. This is *local periodicity.*

## 3. Background and Motivation

Our interest in the c.c.a. $\zeta_t$ grew out of work on similar cyclic systems $\xi_t$ with *random* dynamics, described in [2] and [8]. In those models, with time either discrete or continuous, the type at each site chooses a neighbor at random and eats that neighbor if it can. The main result of [2] deals with one-dimensional (continuous-time) $\xi_t$. It is proved that for $N \geq 5$ the system *fixates.* That is to say, each site attains a final random type (with probability one). If $N = 3$ or 4, on the other hand, $\xi_t$ *clusters,* meaning that contiguous regions of all one type grow arbitrarily large (in distribution). Article [8], an expository piece written for a general audience, describes some Cellular Automaton Machine experiments on the two-dimensional $\xi_t$. To our great surprise, CAM gave compelling evidence that these systems neither cluster nor fixate for *any* value of $N$. Rather, after an initial metastable period of wave droplet formation and growth, the dynamic settles down to a remarkably stable statistical equilibrium consisting of large-scale spiraling wave formations, as illustrated in Plate 6. When $N$ increases, both the duration of the initial metastable phase, and the size of the eventual vortices, grow rapidly. For more details and a few pictures see [8].

In some cases, deterministic cellular automata are easier to analyze than their random counterparts. For instance, some rigorous results for the one-dimensional c.c.a. $\zeta_t$ are derived in [5] and [6]. Reference [5] establishes fixation for $N \geq 5$ and fluctuation for $N = 3, 4$, in direct analogy to the results of [2]. Paper [6] computes the exact asymptotic rate of clustering for the 3-color c.c.a. on $\mathbb{Z}$: as $t \to \infty$,

$$(\dagger) \qquad \text{Average cluster size at time } t \sim \sqrt{\frac{3\pi t}{2}}$$

(with probability one); the corresponding problem for $\xi_t$ is open. We note in passing that the calculation (†) is accomplished by considering the evolution of cluster *edges* as a deterministic particle system, one closely related to the deterministic model analyzed by Harris in [9]. Instead of reflecting, the particles in [6] annihilate upon collision.

We initiated our investigation of the two-dimensional c.c.a. in the hope that they might shed light on the random cyclic systems $\xi_t$. In spite of the fact that the deterministic rules lead to local periodicity and the random rules lead to stable random equilibria, many qualitative and quantitative features of the two dynamics are shared. In the next section we identify several regularity properties of $\zeta_t$ that improve the prospects for detailed rigorous analysis. Such an analysis may, in turn, aid our understanding of the more mysterious behavior of $\xi_t$.

In the general classification of cellular automata, both fixation and local periodicity are instances of "Class 2" behavior, according to the scheme of Wolfram [16],[11]. (It is less clear where *clustering* fits into that scheme.) Local periodic spiral formations are present in two-dimensional cellular automaton models for: atrial fibrillation [10], Belousov-Zhabotinsky chemical reactions [13],[15], neural networks with firing and recovery [3],[12], and related nonlinear wave dynamics. As far as we are aware, there are no substantial rigorous mathematical results for any of these systems. We propose the c.c.a. $\zeta_t$ as relatively tractable prototypes for this important variety of complex spatial phenomena.

## 4. Regularity Properties

Let us now define *debris, droplets, defects*, and *demons*, the principal features of c.c.a. dynamics. First, site $x$ belongs to the *debris* $D_t^0$ at time $t$ if its type $\zeta_t(x)$ differs from that of each of its neighbors by at least 2 (mod $N$). We set $D_t^1 = \mathbb{Z}^2 - D_t^0$, call the connected components of $D_t^1$ *droplets*, and for $x \in D_t^1$, let $D_t^1(x)$ denote the droplet containing $x$. Roughly speaking, the fine-grained regions in Plate 1 correspond to $D_{100}^0$, whereas the open fields comprise $D_{100}^1$. Note that by time 200, in Plate 2, very little debris remains. This reflects the first key property of $\zeta_t$:

$$(\star_1) \qquad D_t^1 \text{ is increasing in } t.$$

Given neighboring sites $x, y$, say that bond $\{x, y\}$ is *open* at time $t$ if $\zeta_t(y) - \zeta_t(x) \in \{-1, 0, 1\}$ (mod $N$). Property $(\star_1)$ follows from the simple observation that an open bond stays open at all subsequent times. Call $\{x, y\}$ *active* at time $t$ if $\zeta_t(y) - \zeta_t(x) \in \{-1, 1\}$ (mod $N$), *inert* if $\zeta_t(y) = \zeta_t(x)$, and *closed* otherwise. A droplet is *active* at time $t$ if it involves at least one active bond; otherwise it is *inert*. Whereas the set of all droplets increases, over time it is quite possible for $D_t^1(x)$ to switch repeatedly between being active and inert.

Next, we introduce the *index* $\iota_t(\ell)$ of $\zeta_t$ restricted to an open loop $\ell$ in $D_t^1$. Let $\ell = z_0, z_1, \ldots, z_n$ be an *open loop* in $D_t^1$ (with $\{z_k, z_{k+1}\}$ an *open bond* for each $k$, and $z_n = z_0$). Let $S_m$ be the unit circle $S^1$ marked with $m$ equidistant vertices, labeled $v_0, \ldots, v_{m-1}$ (counterclockwise). Given $\ell$, take $\phi : S_n \to S_N$ to be the linear extension of the vertex map

$$\phi(v_k) = v_{\zeta_t(z_k)}.$$

Now define $\iota_t(\ell)$ to be the index (winding number) of $\phi$. Figure 1 below shows loops of index 0 and index 1 in sample 8-type configurations; these examples should make the definition clear. A loop in $D_t^1$ with positive index is called a *defect*. We let $D_t^2$ denote the set of all sites that are part of some defect at time $t$, and put $D_t^2(x) = D_t^2 \cap D_t^1(x)$. Of course any defect must involve all $N$ types, so in fact, $D_t^2$ is contained in the set of $D_t^+$ of active droplets at time $t$.

| | | | | | | |
|---|---|---|---|---|---|---|
| 0 | 1 | 2 | | | | |
| 1 | | 3 | 4 | 5 | 6 | |
| 2 | | | | | 7 | An 8-type loop of index 0 |
| 3 | 4 | | | | 6 | |
| | 5 | 6 | 7 | 6 | 5 | |

| | | | | | | |
|---|---|---|---|---|---|---|
| 0 | 1 | 2 | | | | |
| 7 | | 3 | 4 | 5 | 4 | |
| 6 | | | | | 3 | An 8-type loop of index 1 (a defect) |
| 6 | 5 | | | | 2 | |
| | 4 | 5 | 4 | 3 | 2 | |

*Fig. 1.* Sample loops of index 0 and 1 in the $N = 8$ cyclic cellular automaton.

The characteristic c.c.a. spirals, such as those forming at time 200 in Plate 2, indicate the presence of defects. In fact, any loop contained within one of these formations and surrounding its "center" is a defect. The spread of spirals by time 300, evident in Plate 3, reflects a second key property of $\zeta_t$:

$(\star_2)$ $\qquad D_t^2$ is increasing in $t$.

To verify this claim, we make use of the formula

$$\iota(\ell) = N^{-1} \sum_{k=0}^{n-1} \{\zeta_t(z_{k+1}) - \zeta_t(z_k)\},$$

where the terms in brackets are all chosen mod $N$ to be either $-1, 0$, or 1 (since $\ell \subset D_t^1$), and the addition is ordinary (*not* mod $N$). It is easy to check, case by case, that this sum of differences cannot change from time $t$ to $t+1$. Thus, the index of any loop is invariant once that loop is open. In particular, defects are permanent, property ($\star_2$) holds, and any defect present at time $t$ belongs to $D_u^+$ for all $u \geq t$.

The transition of $\zeta_t$ from Stage 3 to Stage 4 is governed by *clocks* and *demons*. A *clock* at time $t$ is a loop $\ell$ such that $\zeta_t(z_{k+1}) - \zeta_t(z_k) \equiv 1$ $(0 \leq k \leq n-1)$. Evidently the size of the loop must be a multiple of $N$. The set $D_t^3$ consists of all $x$ that belong to a clock, or can be reached from one by a path of $+1$ bonds. More precisely, $x \in D_t^3$ provided there exists a path $\mathcal{P} = (y_1, \ldots, y_m)$ of sites such that (i) $\{y_k, y_{k+1}\}$ is a bond for each $k$, (ii) $y_0 = x$ and $y_m$ belongs to a clock at time $t$, and (iii) $\zeta_t(y_{k+1}) - \zeta_t(y_k) \equiv 1$ $(1 \leq k \leq m-1)$. Connected components of $D_t^3$ are called *demons*; we denote the demon containing $x$ by $D_t^3(x)$.

As explained earlier, the onset of local periodicity involves the displacement of "less efficient" spiral formations by demons. This process reflects yet a third key property of $\zeta_t$:

($\star_3$) $\qquad D_t^3$ is increasing in $t$.

Clocks are clearly stable under the c.c.a. dynamic. Since any site connected to a clock by a path $\mathcal{P}$ of $+1$ bonds remains "in phase" with the clock at all later times, ($\star_3$) follows. Such paths are clock-driven "tentacles" of the demon, which cycle at maximal speed, and latch on to neighboring sites unless their types also cycle in $N$ steps.

Note that monotonicity properties ($\star_1$)-($\star_3$) apply to any c.c.a., either on $\mathbb{Z}^2$ or a finite lattice of width $L$, starting from any initial configuration, either deterministic or random. Letting $T_t^k(x) = \inf\{t \geq 0 : D_t^k(x) \neq \emptyset\}$ $(k = 1, 2, 3)$, it follows that

$$T_t^1(x) \leq T_t^2(x) \leq T_t^3(x)$$

in full generality. As we shall now see, existence of a single clock ensures periodicity at every $x$.

## 5. Local Periodicity

Say that a cellular automaton $\zeta_t$ with random initial state is *locally periodic* if

(i) for each $x$, $\zeta_t(x)$ is eventually periodic in $t$ with probability one, and
(ii) $\inf_{t,x\neq y,a,b} P(\zeta_t(x) = a, \zeta_t(y) = b) > 0.$

Condition (i) asserts that the type at each site settles into a periodic orbit eventually (with period $> 1$). Condition (ii) requires that self-organization remain local: correlations between distinct sites must be uniformly controlled in space and time. In particular, the latter condition distinguishes local periodicity from *clustering*, where for each $x, y$,

$$\lim_{t\to\infty} P(\zeta_t(x) = \zeta_t(y)) = 1.$$

We now present a proof that, starting from a random initial configuration $\zeta_0$, the c.c.a. on $\mathbb{Z}^d$, $d \geq 2$, is always locally periodic, with minimal period at every site, no matter how large the number of colors $N$. Thus the limiting behavior is quite different from that of one-dimensional cyclic systems ([2],[5],[6]), which either fixate or cluster.

Verification of (i) is based on the following simple observation.

**Lemma.** *If there is a clock anywhere in $\zeta_0$ (or if a clock forms at some later time $t$), then every site $x$ is eventually periodic with period $N$.*

*Proof.* Let $C$ be the set of sites that have period $N$ eventually in $t$. $C \neq \emptyset$, since the c.c.a. contains a clock by hypothesis. Suppose $x \notin C$, $y \in C$, $\|y - x\| = 1$. Then the difference $\zeta_t(y) - \zeta_t(x) \in \{0, \ldots, N-1\}$ (mod $N$) is eventually non-increasing. Hence this difference is eventually constant. But then $x \in C$, a contradiction. We conclude that every site $x$ belongs to $C$, as claimed. (We note that in CAM experiments with $N$ large, the eventual constant difference is almost invariably 1 or 0. In the former case, $x$ and $y$ are part of the same demon; otherwise bond $\{x, y\}$ almost always belongs to the boundary between two distinct demons.)

Condition (ii) is even easier to check. Given any time $t$, distinct sites $x$ and $y$, and types $a$ and $b$, the event in question occurs provided that $\zeta_0$

contains a pair of disjoint clocks through $x$ and $y$ with appropriate phases. The probability of this latter occurrence is at least

$$\left(\frac{1}{N}\right)^{4N} > 0.$$

## 6. Spirals: what, where, when, and why?

Although the argument just given determines the limiting behavior of any c.c.a. on $\mathbb{Z}^2$, it does not capture the true nature of the dynamics. CAM experiments clearly indicate that the first demon to reach a typical site $x$ is *not* present in the initial state, but "pops up" at a later time. In other words, defects in $\zeta_0$, used to prove local periodicity of the infinite system, occur on too large a scale.

Many intriguing problems concerning c.c.a. evolution are suggested by Plates 1–5. In this section we make some empirical observations that shed light on perhaps the four most fascinating questions:

*What are the characteristic c.c.a. spirals, and where, when and why do they pop up?*

*What?—*

When $N$ is small, the spirals of $\zeta_t$ have noticeable "glitches." As these coherent structures form, there is enough variation in the environment that occasional "errors" occur. But as $N$ increases, the individual spirals grow in final size *and* exhibit fewer errors. What is an *ideal spiral*? Here, at least, is an example. Suppose $N = 14$. Let $\ell$ be a "convex" loop of length 14 in $\mathbb{Z}^2$, say the boundary of a $5 \times 4$ rectangular array of sites. Specify $\zeta$ on $\ell$ to be a clock. Choose types for the interior of the array so that all interior sites are connected to the clock by a path of $+1$ edges. Now define $\zeta$ outside of $\ell$ by the formula:

$$(\S) \quad \zeta(x) = \zeta(x^*) + \rho(x, x^*), \quad \text{where} \quad \rho(x, x^*) = \min\{\rho(x, y) : y \in \ell\},$$

and $\rho(x, y)$ denotes the length of a shortest path in $\mathbb{Z}^2$ between $x$ and $y$. We invite the reader to color a piece of graph paper according to (§) in order to confirm the connection with Plates 2–4. Recipe (§) also generates representative spirals when $\zeta$ on $\ell$ is a defect but not a clock. If $\ell$ is not convex, then, roughly speaking, the recipe should be applied to a suitable "convexification" of $\ell$.

*Where?*—

CAM experiments strongly suggest that spirals can only pop up on the interfaces between droplets and debris. In fact, this is a topological necessity. Recall that any loop of sites surrounding the center of a spiral has index at least one, and that the index of any loop is invariant once it is open. It follows that if a defect $\ell$ forms at time $t > 0$ (i.e., $\ell \not\subset D^1_{t-1}, \ell \subset D^1_t$), then $\ell$ must involve at least one closed bond at time $t-1$. Either some site of $\ell$ belongs to the debris, or $\ell$ involves two distinct droplets, at that previous time.

*When?*—

Suppose we lay down a uniform random distribution of $N$ types on $\mathbb{Z}^2$, and then construct a starting configuration $\zeta_0$ by modifying the values on a suitable loop $\ell$ to produce a clock. For instance, if $N = 14$, we could change the values on the boundary of a $5 \times 4$ array of sites. What happens when the c.c.a. starts from such an initial state? With the aid of a microcomputer, the enterprising reader can verify that $\zeta_t$ does *not* give rise to well-formed spirals. A roughly circular disturbance spreads out from the clock, but it is fragmented and incoherent, with remarkably little resemblance to the spirals encountered before. This is a striking indication of the fact, already mentioned, that characteristic c.c.a. spirals are produced by the dynamic, not the initial noise. Aided by CAM we discover that at typical sites $x$, all of the random times $T^k(x)$ increase rapidly with $N$. In particular, since the distance between critical droplets becomes very large, and since the formation of defects seems typically to involve interaction between distinct droplets, spirals only tend to pop up after most of the debris has been displaced.

*Why?*—

How does the c.c.a. evolve if $\zeta_0$ consists of a clock on a background of all one type? We encourage the reader to solve this riddle, either by hand or by computer. Again, surprisingly, waves emanating from the clock do *not* make a spiral. So what *is* the proper environment for spiral creation? In an attempt to solve this mystery we examined a few trajectories of the 18-color c.c.a. (with $L = 256$), carefully watching for the appearance of the first spiral.

In one typical instance the first spiral "popped up" at about time 800. It contained a clock $\ell$ consisting of 18 sites, and by time 850 had formed a perfect spiral satisfying (§) without error for a distance of about 50 bonds in all directions. We then worked *backwards* through the (deterministic) history of $\zeta_t$, in order to discover the time

$$T^2 = \min_x T^2(x)$$

of the first defect anywhere in the $256 \times 256$ lattice. In this case we determined $T^2 = 727$. Many different defects formed at that time, since various routes across numerous wide bands of solid type were possible. As shown in Plate 7, all these routes had to travel through a narrow passage of debris, leading to a "pivotal" bond that completed the loops by opening at time $T^2$. The first defects were remarkably long, involving over 100 sites that had previously belonged to three different large droplets. At time 727 there was no real indication of a spiral. After $T^2$, however, shorter interior defects rapidly formed, feeding on remnants of debris, until the clock appeared. This process of consolidation, it would seem, created the proper environment for a perfect spiral.

## 7. Additional challenging problems

Our preliminary investigation of the c.c.a. rule on $T^2$ suggests many challenging mathematical problems. In this final section we mention a few.

The early stages of c.c.a. dynamics are *metastable*. To understand their quantitative aspects one should study finite systems of size $L^2$, as $N$ and $L$ increase appropriately, along the lines of the asymptotic analysis for *bootstrap percolation* presented in [1]. However the lack of monotonicity and manifest complexity of cyclic systems make the corresponding agenda more challenging. We have therefore collected some empirical data as a first step. Here we present only a sampling of our findings; a more complete account will appear in [7].

Figures 2 and 3 show the frequencies of active and inert bonds over time in the $N$-color c.c.a. on a $256 \times 256$ lattice with wrap-around, for $N$ ranging from 12 to 16. Each plot tracks the average values obtained from 50 trajectories. In Figure 2, first note the rapid initial decrease. This corresponds to Stage 1, when most creatures run out of food. But droplet growth signals the onset of Stage 2 and causes the density of active bonds to increase. The density then continues increasing to a limit: roughly, the probability that adjacent sites belong to a common demon. In Figure 3, the proportion of inert bonds at first increases very rapidly, as initially active bonds form small

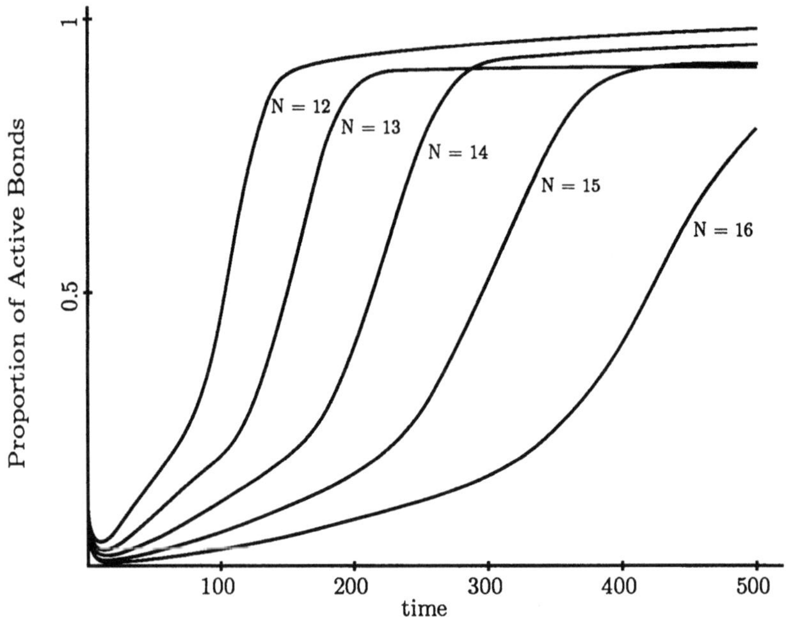


*Fig. 2.* Density of active bonds in c.c.a. (L=256)

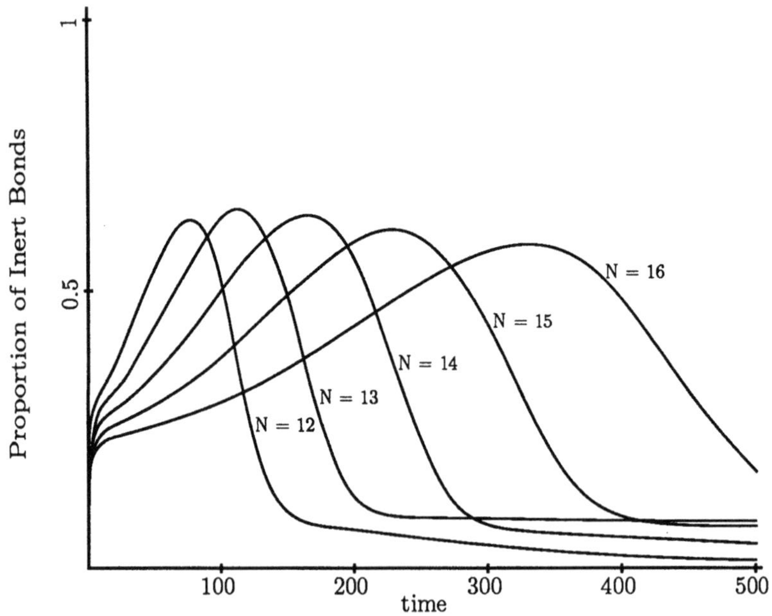


*Fig. 3.* Proportion of inert bonds in c.c.a. (L=256)

inert clusters. There follows a longer period of slower but steady increase, signaling the presence of waves of solid type in the growing droplets. At the end of Stage 2 a maximum is reached, and then the freqency begins to dip as bonds are reactivated by defects and demons. The eventual limit is, roughly, the (very small but positive) probability that adjacent sites belong to different demons.

We note that the sum of the corresponding frequencies in Figures 2 and 3 (i.e., the density of open bonds) increases monotonically in accordance with ($\star_1$). In our data the limit of the sum is very close to 1; in particular, none of our samples fixated. Evidently, the theoretical probability of fixation for $L = 256$ and $N \leq 16$ is extremely small. With $L$ fixed and $N$ increasing, fixation continues to be very rare until an apparent critical value $N_c(L)$ when critical droplets are too rare to appear in the $L$-system. Although we have not checked, we expect that $N_c(256)$ lies in the low 20's. For values of $N$ beyond that point, the curves corresponding to Figure 2 will appear, for all practical purposes, to decrease exponentially to 0. As beautifully illustrated in [1], this "pseudo phase transition" should become more pronounced as $L$ increases, but $N_c(L)$ will increase *very slowly* to $+\infty$.

At what rate should $N_c(L)$ increase with $L$? Based in part on weak analogy with bootstrap percolation, and in part on heuristic arguments, we make the following conjecture, which suggests *logarithmic* growth:

**Conjecture.** *Let $p(N, L)$ denote the probability that the $N$-color c.c.a. on an $L \times L$ lattice (with wrap-around) fixates. Then there are constants c and $C$ $(0 < c < C < \infty)$ such that*

$$\begin{aligned} \lim_{N\to\infty} p(N, L_N) &= 1 \qquad \textit{whenever } L_N < e^{cN}, \\ &= 0 \qquad \textit{whenever } L_N > e^{CN}. \end{aligned}$$

We suspect that the first assertion of the conjecture may follow from percolation considerations. To establish the second claim, one will need a better understanding of critical droplets and how they grow. A *localized* c.c.a. rule may prove helpful in this connection. Let $\zeta_0^0$ be a uniform random configuration of $N$ types on $\mathbb{Z}^2$ as before. Inductively define $\zeta_t^0$ and a sequence $A_t$ of finite subsets of $\mathbb{Z}^2$ by: $A_0 = \{0\}$ (the origin), and

$$(\circ) \qquad \begin{aligned} \zeta_{t+1}^0(x) &= \zeta_t^0(y) \quad \text{if } \zeta_t^0(y) - \zeta_t^0(x) = 1 \bmod N \text{ for some } y \text{ such that} \\ &\qquad\qquad \|y - x\| = 1, \text{ and } (x \in A_t \text{ or } y \in A_t), \\ &= \zeta_t^0(x) \quad \text{otherwise.} \\ A_{t+1} &= A_t \cup \{x : \zeta_{t+1}^0(x) \neq \zeta_t^0(x)\}. \end{aligned}$$

Roughly speaking, algorithm (o) isolates the droplet at the origin. This droplet will stop growing with overwhelming probability when $N$ is large. But if it reaches a certain critical size, then it almost invariably survives forever, acquiring an asymptotic shape as it spreads out at a linear rate. The three large droplets in Plate 5 evolve in essentially the same manner until they collide. This localized rule is interesting in its own right, as a deterministic dynamic with behavior reminiscent of stochastic growth models.

Finally, most of our discussion also applies to corresponding dynamics in three or more dimensions. Winfree and Strogatz [14] have made a rather thorough study of the locally periodic formations that can arise from systems of interacting cyclic waves in three-dimensional space. We plan to investigate $\mathbb{Z}^3$ counterparts of the ideal spirals in Plates 2–3, and compare our findings with those of [14]. Plate 8 shows a c.c.a. configuration for a rule having 12 nearest neighbors instead of 4.

## References


[1] Aizenman, M. and Lebowitz, J., *Metastability effects in bootstrap percolation*, J. Phys. A: Math. Gen. **21** (1988), pp. 3801–3813.

[2] Bramson, M. and Griffeath, D., *Flux and fixation in cyclic particle systems*, Annals of Probability **17** (1989), pp. 26–45.

[3] Dewdney, A.K., *Computer Recreations*, Scientific American **August** (1988), pp. 104–107.

[4] Dewdney, A.K., *Computer Recreations*, Scientific American **August** (1989), pp. 102–105.

[5] Fisch, R., *The one-dimensional cyclic cellular automaton: a system with deterministic dynamics which emulates an interacting particle system with stochastic dynamics*, J. Theoretical Probability **3** (1990), pp. 311–338.

[6] Fisch, R., *Clustering in the one-dimensional 3-color cyclic cellular automaton*, Ann. Probab. (1989) (to appear).

[7] Fisch, R., Gravner, J. and Griffeath, D., *Threshold-range scaling of excitable cellular automata*, Statistics and Computing (to appear).

[8] Griffeath, D., *Cyclic random competition: a case history in experimental mathematics*, In "Computers and Mathematics," ed. J. Barwise, A.M.S. Notices **December** (1988), pp. 1472–1480.

[9] Harris, T.E., *Diffusion with "collisions" between particles*, J. Appl. Probability **2** (1965), pp. 323–338.

Plate 1

Plate 2

Plate 3

Plate 4

CYCLIC CELLULAR AUTOMATA IN TWO DIMENSIONS

Plate 5

Plate 6

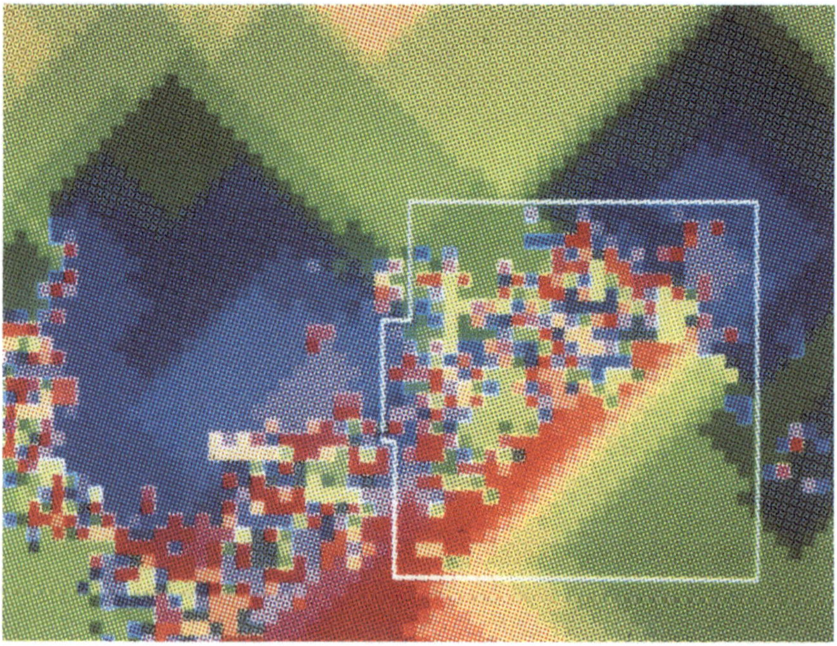

Plate 7

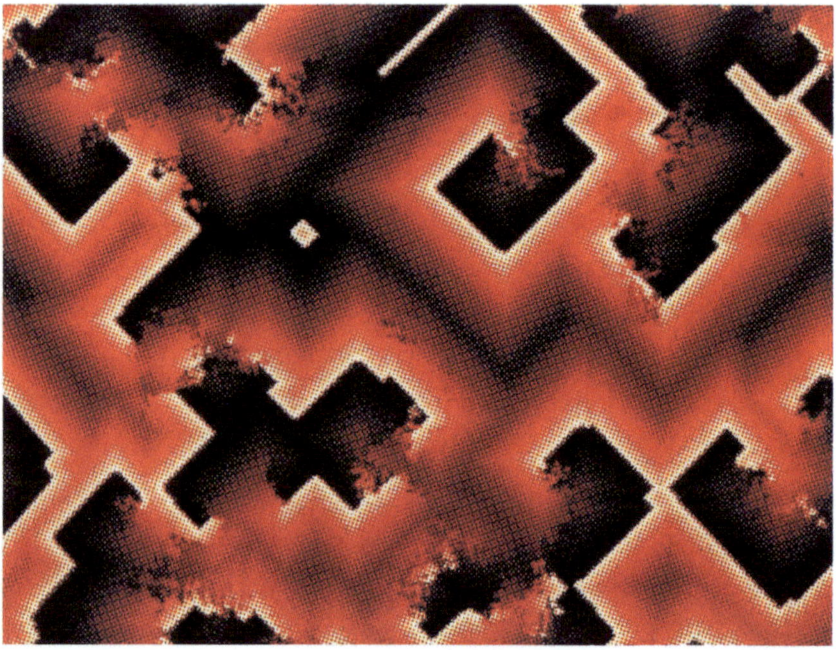

Plate 8

[10] Moe, G.K., Rheinboldt, W.C. and Abildskov, J.A., *A computer model of atrial fibrillation*, American Heart J. **67** (1964), pp. 200–220.

[11] Packard, N. and Wolfram, S., *Two-dimensional cellular automata*, J. Stat. Phys. **38** (1985), pp. 901–946.

[12] Toffoli, T. and Margolus, N., "Cellular Automata Machines," M.I.T. Press, Cambridge, Massachusetts, 1987.

[13] Winfree, A.T., *Rotating chemical reactions*, Scientific American **230 June** (1974), pp. 82–95.

[14] Winfree, A.T. and Strogatz, S.H., *Organizing centres for three-dimensional chemical waves*, Nature **311 no. 5987** (1984), pp. 611–615.

[15] Winfree, A.T., Winfree, E.M. and Seifert, H., *Organizing centers in a cellular excitable medium*, Physica **17D** (1985), pp. 109–115.

[16] Wolfram, S., *Universality and complexity in cellular automata.*, Physica **10D** (1985), pp. 1–35.

Research supported in part by NSF Grant DMS-87-0083
Electronic mail: griffeat@math.wisc.edu

Robert Fisch
Department of Mathematics
Colby College
Waterville, ME 04901

Janko Gravner
Department of Mathematics
University of Wisconsin
Madison WI 53706

David Griffeath
Department of Mathematics
University of Wisconsin
Madison WI 53706

# Stochastic Flows

# Statistical Equilibrium and Two-Point Motion For a Stochastic Flow of Diffeomorphisms

Peter H. Baxendale

## 1. Introduction

Let $M$ be a connected, smooth, compact manifold of dimension $N$, and consider the following Stratonovich stochastic differential equation on $M$

$$\text{(1.1)} \qquad \begin{cases} d\xi_t(x) &= \sum_{k=1}^{d} X_k(\xi_t(x)) \circ dW_t^k + X_0(\xi_t(x))\, dt \\ \xi_0(x) &= x. \end{cases}$$

Here $X_0, X_1, \ldots X_d$ are smooth vector fields on $M$ and $\{W_t^k : t \geq 0\}, 1 \leq k \leq d$, are independent real valued Brownian motions on some probability space $(\Omega, \mathcal{F}, \mathbb{P})$. For each $x \in M$ (1.1) has a (unique) strong solution which is a diffusion on $M$ with generator

$$\text{(1.2)} \qquad L = \tfrac{1}{2} \sum_{k=1}^{d} X_k^2 + X_0.$$

Let $\{\xi_t : t \geq 0\}$ denote the stochastic flow of diffeomorphisms of $M$ generated by (1.1). Thus $\{\xi_t : t \geq 0\}$ is a process defined on $(\Omega, \mathcal{F}, \mathbb{P})$ with values in Diff($M$), the group of smooth diffeomorphisms of $M$, such that for all $x \in M$ the one-point motion $\{\xi_t(x) : t \geq 0\}$ coincides almost-surely with the strong solution of (1.1). For results on existence and regularity properties for the process see for example Elworthy [7], Ikeda and Watanabe [12], Kunita [14], or Rogers and Williams [19].

We shall be concerned here with questions concerning the behavior of the stochastic flow $\{\xi_t : t \geq 0\}$ as $t \to \infty$. This topic has been studied recently by various authors, not only for stochastic flows as defined above but also for more general versions of the definition (where possibly discrete time and/or non-compact $M$ are allowed). In addition to work cited later in this paper we refer the reader to work of Harris [9], Baxendale and Harris [3], Kifer [13], and Darling [6]. For a more detailed introduction to the class of stochastic flows discussed here see Baxendale [2].

The results presented in this paper concern the following:

**A) Stability of solutions.**

For distinct $x, y \in M$ what happens to $dist(\xi_t(x), \xi_t(y))$ as $t \to \infty$ ? (Here *dist* denotes the geodesic distance on $M$ corresponding to a choice of smooth Riemannian structure on $M$; the compactness of $M$ implies that *dist* is well defined up to uniform equivalence.) In other words, how sensitive is the stochastic flow to perturbations of initial conditions?

**B) The two-point motion.**

Let $\Delta$ denote the diagonal in $M \times M$. Then $\hat{M} \equiv M \times M \setminus \Delta$ is an invariant set for the two-point motion $\{(\xi_t(x), \xi_t(y)) : t \geq 0\}$ on $M \times M$, and we shall concentrate attention on the two point motion on $\hat{M}$. (On $\Delta$, the two-point motion collapses down to a copy of the one-point motion.) Since $\hat{M}$ is non-compact, questions of transience or recurrence of the two-point motion on $\hat{M}$ arise. Notice that the statement that $dist(\xi_t(x), \xi_t(y)) \to 0$ as $t \to \infty$ can be intepreted simultaneously as a statement of stability of the stochastic flow and also as a statement about transience of the two-point motion on $\hat{M}$. The closely related question of the existence of a stationary probability measure for the two point motion on $\hat{M}$ will provide a link with the next topic.

**C) Statistical equilibrium.**

Let $\rho$ be a stationary probability measure for the one-point motion on $M$, and write $\rho_t = \xi_t \rho$ for $t \geq 0$. Then $\{\rho_t : t \geq 0\}$ is a random process in $\mathcal{P}(M)$, the space of Borel probability measures on $M$. Notice that the assumption that $\rho$ is stationary for the one-point motion is equivalent to the equality $\mathsf{E}(\rho_t) = \rho$ for all $t \geq 0$. This is a weaker assertion than saying that $\rho$ is invariant under the stochastic flow, i.e. that $\rho_t = \rho$ for all $t \geq 0$ almost surely. (This is a distinction which does not occur in non-random dynamical systems.) It has been shown (by Le Jan [16], see also Young [21]) that, as a family of $\mathcal{P}(M)$ valued random variables, $\rho_t$ converges in distribution as $t \to \infty$. The limiting distribution $Q$, say, is called the *statistical equilibrium*. It proves a stationary family of random probability measures for the stochastic flow. Heuristically, a typical $\nu \in \mathcal{P}(M)$ chosen according to law $Q$ is the typical distribution on $M$ obtained by running the stochastic flow for a very long time when started with initial distribution $\rho$. Our aim here is to describe $Q$ as fully as possible.

In Section 2 we present details concerning statistical equilibrium and the two-point motion. We show that the second moment of the statistical equilibrium is stationary for the two-point motion on $M \times M$. This establishes a link between questions $B$ and $C$ (see Propositions 2.7 and 2.8 for details). After briefly introducing the Lyapunov exponents $\lambda_1 \geq \lambda_2 \geq \ldots \geq \lambda_N$ in Section 3, we present the main results of the paper in Sections 4 and 5. These show how the behavior of the two-point motion on $\hat{M}$ and the nature of the statistical equilibrium are influenced by the value of $\lambda_1$, the top Lyapunov exponent. Section 4 contains mostly known results due to Le Jan [16], Baxendale and Stroock [4], and Ledrappier and Young [15] for the cases $\lambda_1 < 0$ and $\lambda_1 > 0$, while Section 5 contains new results for the case $\lambda_1 = 0$.

The main results on the nature of the statistical equilibrium are given in Theorems 4.10, 4.12, 4.18 and Corollaries 4.16 and 5.12. Together they say that, if the vector fields $X_0, X_1, \ldots X_d$ satisfy suitable non-degeneracy conditions, $Q$ consists of unit point measures when $\lambda_1 \leq 0$, while if $\lambda_1 > 0$ then $Q$ consists of measures of Hausdorff dimension equal to the Lyapunov dimension determined by the Lyapunov exponents $\lambda_1 \geq \ldots \geq \lambda_N$.

Results on the two-point motion (and hence on the stability question $A$) are given in Theorems 4.10, 4.12, 4.14, 5.8 and Corollary 5.4. Those are bought together in Section 6 where it is shown, again under appropriate non-degeneracy conditions, that the two-point motion on $\hat{M}$ is positive recurrent, null recurrent, or transient according as $\lambda_1$ is positive, zero, or negative.

## 2. Statistical Equilibrium and the Two-Point Motion

In the introduction statistical equilibrium was defined as the distribution limit of $\rho_t$ as $t \to \infty$. To show that this limit exists, however, it is convenient to think of it as an almost sure limit obtained by starting the stochastic flow with distribution $\rho$ in the far distant past. This idea is made precise as follows. By extending the probability space if necessary we may assume that the Brownian motions appearing in (1.1) are the restrictions to positive time of independent two-sided Brownian motions $\{W_t^k : t \in R\}, 1 \leq k \leq d$. Consider the random diffeomorphisms $\{\xi_{s,t} : -\infty < s \leq t < \infty\}$ determined by

$$\xi_{s,v}(x) = x + \sum_{k=1}^{d} \int_s^v X_k(\xi_{s,v}(x)) \circ dW_u^k + \int_s^v X_0(\xi_{s,v}(x))\, du \tag{2.1}$$

for $s \leq v \leq t$. We may think of $\xi_{s,t}$ as describing the random evolution of the state space $M$ from time $s$ to time $t$. For $s > t$ write $\xi_{s,t} = (\xi_{t,s})^{-1}$. Notice that $\{\xi_{0,t} : t \geq 0\}$ is the original stochastic flow determine by (1.1).

Let $\theta_s : \Omega \to \Omega$ denote the canonical shift, so that $W_t^k(\theta_s\omega) = W_{t+s}^k(\omega) - W_s^k(\omega)$ for $\omega \in \Omega, s, t \in \mathbb{R}$. Then, writing $\xi_{s,t,\omega}$ to denote the dependence of $\xi_{s,t}$ on $\omega \in \Omega$, we have

$$\xi_{t,u,\omega} \circ \xi_{s,t,\omega} = \xi_{s,u,\omega} \tag{2.2}$$

$$\xi_{s,t,\theta_u\omega} = \xi_{s+u,t+u,\omega} \tag{2.3}$$

for $\omega \in \Omega, s, t, u \in \mathbb{R}$. If $\mathcal{F}_s^t = \sigma\left\{W_v^k - W_u^k : s \leq u \leq v \leq t, 1 \leq k \leq d\right\}$ then $\xi_{s,t}$ is $\mathcal{F}_s^t$-measurable and the $\mathcal{F}_s^t$ corresponding to disjoint intervals are independent.

For $t > 0$, $\xi_{-t,0}\rho$ represents the random measure obtained at time 0 when the system is started at time $-t$ with distribution $\rho$. Since $\xi_{-t,0,\omega} = \xi_{0,t,\theta_{-t}\omega}$ then, for each fixed $t$, $\xi_{-t,0}$ and $\xi_{0,t}$ have the same distribution. (Beware: the processes $\{\xi_{-t,0} : t \geq 0\}$ and $\{\xi_{0,t} : t \geq 0\}$ have their increments composed on different sides and hence usually evolve according to different laws.) This implies that $\xi_{-t,0}\rho$ and $\rho_t$ have the same distribution in $\mathcal{P}(M)$, and in particular $\mathbb{E}(\xi_{-t,0}\rho) = \mathbb{E}(\rho_t) = \rho$ for all $t \geq 0$. This equation, together with (2.2) and the independence of $\mathcal{F}_{-s-t}^{-t}$ and $\mathcal{F}_{-t}^0$, implies that $\left\{\xi_{-t,0}\rho; \mathcal{F}_{-t}^0\right\}_{t\geq 0}$ is a measure valued martingale. It follows that for $\mathbb{P}$-almost all $\omega \in \Omega$

$$\xi_{-t,0,\omega}\rho \to \mu_\omega \qquad \text{as } t \to \infty \tag{2.4}$$

for some $\mu_\omega \in \mathcal{P}(M)$. (The convergence is in the topology of weak convergence in $\mathcal{P}(M)$.) Details of this argument are given in Le Jan [16] and Young [21] in the discrete time case; the extension to continuous time is automatic. Since $\xi_{-t,0}\rho$ and $\rho_t$ have the same distribution for each $t \geq 0$, then $\rho_t \to \mu$ in distribution as $t \to \infty$. Hence the statistical equilibrium $Q$ as defined in the introduction is the image of $\mathbb{P}$ under the $\mathcal{F}_{-\infty}^0$ measurable mapping $\Omega \to \mathcal{P}(M)$ given by $\omega \mapsto \mu_\omega$. In this paper we shall usually refer to statistical equilibrium in terms of the family $\{\mu_\omega : \omega \in \Omega\}$, although any statement about $\mu_\omega$ where $\omega$ has law $\mathbb{P}$ can equally well be made about $v \in \mathcal{P}(M)$ where $v$ has law $Q$. Notice that $\mu_\omega$ is measurable with respect to $\mathcal{F}_{-\infty}^0$ and hence is independent of the original positive time stochastic flow $\{\xi_t : t \geq 0\}$.

Using the definition (2.4) of $\mu_\omega$ together with (2.2) and (2.3) we obtain for all $t \in \mathbb{R}$

$$\xi_{0,t,\omega}\mu_\omega = \mu_{\theta_t\omega} \quad \text{for } \mathbb{P} - \text{almost all } \omega \in \Omega. \tag{2.5}$$

This implies that the family $\{\mu_\omega : \omega \in \Omega\}$ is stationary under the stochastic flow. Let $\{\Theta_t : t \in R\}$ denote the one-parameter group of transformations of $\Omega \times M$ given by $\Theta_t(\omega, x) = (\theta_t\omega, \xi_{0,t,\omega}(x))$. If $\bar{\mathbb{P}}$ denotes the probability measure on $(\Omega \times M, \mathcal{F} \times B(M))$ with marginal $\mathbb{P}$ and regular conditional probability distribution $\{\mu_\omega : \omega \in \Omega\}$, then (2.5) is equivalent to the statement that $\Theta_t$ is $\bar{\mathbb{P}}$-measure preserving. In fact, since $\Theta_t(\mathbb{P} \times \rho)$ has marginal $\mathbb{P}$ and regular conditional probability distribution $\{\xi_{-t,0,\omega}\rho : \omega \in \Omega\}$ then (2.4) can be restated as $\lim_{t\to\infty} \Theta_t(\mathbb{P} \times \rho) = \bar{\mathbb{P}}$.

The previous paragraph was written in terms of the extended flow

$$\{\xi_{s,t} : -\infty < s \le t < \infty\}.$$

We regain results for the original stochastic flow $\{\xi_t : t \ge 0\}$ by taking conditional expectations with respect to $\mathcal{F}_0^\infty$. Since $\mathbb{E}\{\xi_{-t,0,\omega}\rho|\mathcal{F}_0^\infty\} = \mathbb{E}(\xi_{-t,0,\omega}\rho)$ $=\rho$, then $\mathbb{E}\left(\bar{\mathbb{P}}|\mathcal{F}_0^\infty\right) = \mathbb{P} \times \rho$ and we see that $\mathbb{P} \times \rho$ is invariant for $\{\Theta_t : t \ge 0\}$ on $(\Omega \times M, \mathcal{F}_0^\infty \times B(M))$. In this case the stationarity result (2.5) implies that if $v \in \mathcal{P}(M)$ is chosen with law $Q$ and $\xi_t$ is chosen with the law for the stochastic flow, independently of $v$, then $\xi_t v$ has law $Q$ (for $t \ge 0$).

The figures below are taken from a computer simulation of one single evolution of the stochastic flow studied in Section 5 of Baxendale [2]. Here $M = (\mathbb{R}/2\pi\mathbb{Z})^2$, the two-dimensional torus, so each square shown should have its edges identified. Using coordinates $x = \left(x^1, x^2\right)$ we take $X_0 \equiv 0$ and

$$\begin{aligned} X_1(x) &= \sin x^1 (\cos\alpha, \sin\alpha) \\ X_2(x) &= \cos x^1 (\cos\alpha, \sin\alpha) \\ X_3(x) &= \sin x^2 (-\sin\alpha, \cos\alpha) \\ X_4(x) &= \cos x^2 (-\sin\alpha, \cos\alpha). \end{aligned}$$

We show a simulation with $\alpha = \frac{70\pi}{180}$. The one-point motion is Brownian motion on $M$, so that $\rho$ is normalized Lebesque measure. The figures show 400 points in their original array at time 0 together with their images under the flow at times 200, 200.5 and 201. To the extent that 400 points arranged

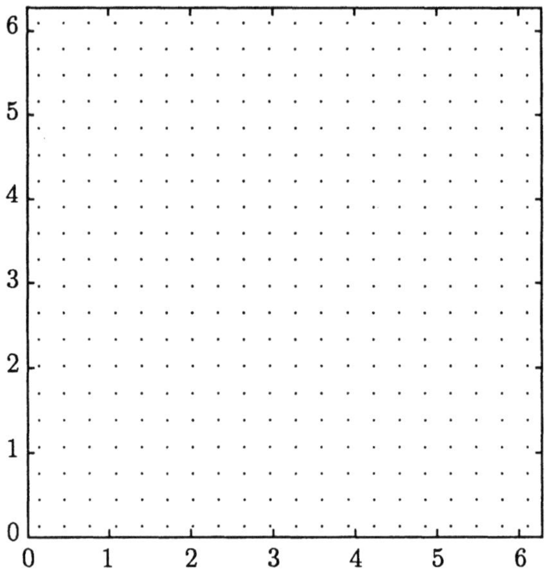


Figure 1: $t = 0$

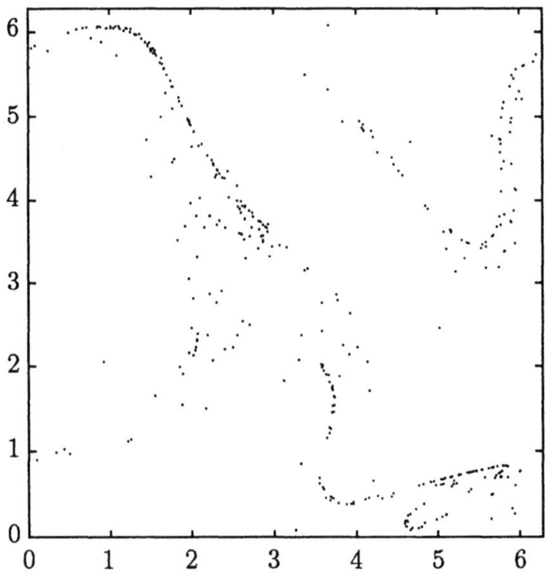


Figure 2: $t = 200$

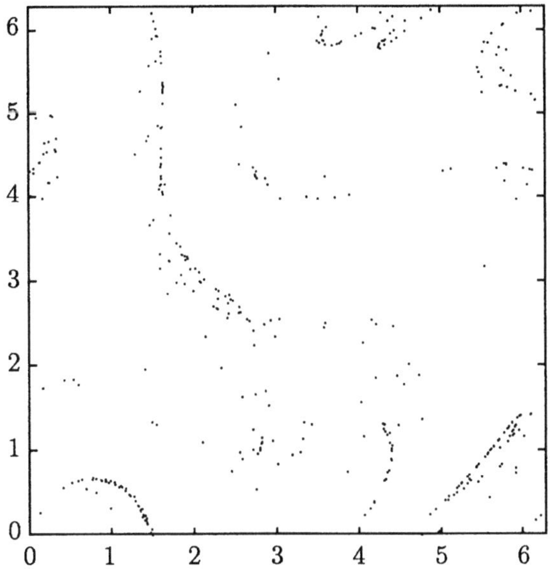


Figure 3: t = 200.5

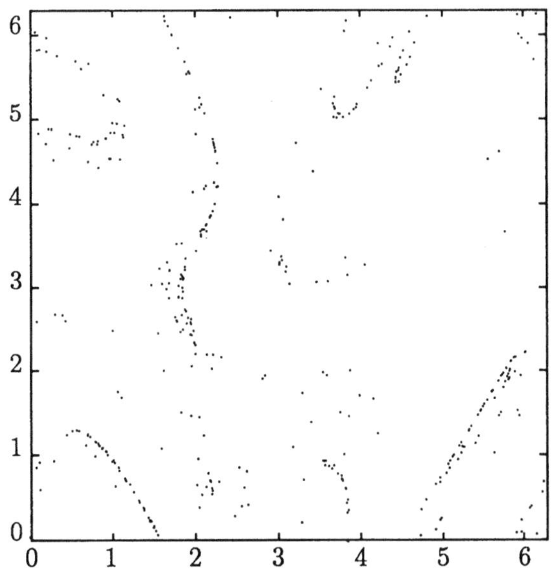


Figure 4: t = 201

in a $20 \times 20$ grid approximate $\rho$, and to the extent that $t = 200$ approximates $t \to \infty$, the later figures also suggest a reason why the $\rho_t$ will not usually converge in a almost sure sense; a typical point chosen with distribution $\rho_t$ will continue to diffuse around $M$ according to the generator $L$ as time proceeds.

Since $\mathbb{E}(\xi_{-t,0,\omega}\rho) = \rho$ for all $t \geq 0$ then $\mathbb{E}(\mu_\omega) = \rho$. There is a similar result relating the second moment of the statistical equilibrium with the two-point motion $\{(\xi_t(x), \xi_t(y)) : t \geq 0\}$ on $M \times M$. Let $\{P_t : t \geq 0\}$ denote the Markov semigroup acting on $C(M \times M)$ associated with the two-point motion and let $\{P_t^* : t \geq 0\}$ denote the adjoint semi-group acting on measures on $M \times M$.

**(2.6) Proposition.**

(i) *$P_t^*(\rho \times \rho)$ converges weakly to $\mathbb{E}(\mu_\omega \times \mu_\omega)$ as $t \to \infty$.*
(ii) *$\mathbb{E}(\mu_\omega \times \mu_\omega)$ is a stationary probability measure for the two-point motion on $M \times M$.*

*Proof.* Let $f \in C(M \times M)$. Since $\xi_{-t,0,\omega}\rho \to \mu_\omega$ weakly in $\mathcal{P}(M)$ then $\xi_{-t,0,\omega}\rho \times \xi_{-t,0,\omega}\rho \to \mu_\omega \times \mu_\omega$ weakly in $\mathcal{P}(M \times M)$. That is

$$\int_{M\times M} f(\xi_{-t,0,\omega}(x), \xi_{-t,0,\omega}(y))\, d(\rho \times \rho)(x,y) \to \int_{M\times M} f(u,v)\, d(\mu_\omega \times \mu_\omega)(u,v)$$

as $t \to \infty$ for $\mathbb{P}$-almost every $\omega \in \Omega$. Taking expectations and using the dominated convergence theorem we have

$$\int_{M\times M} (P_t f)\, d(\rho \times \rho) \to \mathbb{E} \int_{M\times M} f\, d(\mu_\omega \times \mu_\omega)$$

as $t \to \infty$. In other words

$$\int f\, d(P_t^*(\rho \times \rho)) \to \int f\, d(\mathbb{E}(\mu_\omega \times \mu_\omega))$$

as $t \to \infty$. This is true for all $f \in C(M \times M)$, so (i) is proved. The result (ii) now follows from (i) (or alternatively (ii) can be proved directly from formula (2.5)). Q.E.D.

The following two results show how the mass assigned to $\hat{M}$ by the second moment of the statistical equilibrium affects both the two-point motion on $\hat{M}$ and the typical $\mu_\omega$ itself. For convenience we write $\bar{\mu} = \mathbb{E}\left(\mu_\omega \times \mu_\omega\right), r_t(x,y) = dist\left(\xi_t(x), \xi_t(y)\right)$, and $U_\delta = \left\{(x,y) \in \hat{M} : dist(x,y) > \delta\right\}$.

**(2.7) Proposition.** *Assume* $\bar{\mu}(\hat{M}) = 0$.

(i) *There exist (non-random) times* $t_n \to \infty$ *such that, for* $(\rho \times \rho)$*-almost every* $(x,y) \in M \times M$, $r_{t_n}(x,y) \to 0$ *in probability as* $n \to \infty$.
(ii) *Every stationary probability measure for the two-point motion on* $\hat{M}$ *is singular with respect to* $\rho \times \rho$.
(iii) $\mu_\omega$ *consists of a single atom almost surely.*

*Proof.*

(i) By assumption and Proposition 2.6(i), for all $\delta > 0$

$$\lim_{t \to \infty} P_t^* \left(\rho \times \rho\right)\left(U_\delta\right) = 0,$$

i.e.

$$\lim_{t \to \infty} \int_{M \times M} \mathbb{P}\left(r_t(x,y) > \delta\right) d\left(\rho \times \rho\right)(x,y) = 0.$$

In other words, $\mathbb{P}\left(r_t(x,y) > \delta\right) \to 0$ as $t \to \infty$ in $L^1(\rho \times \rho)$. Hence for each $\delta > 0$ there exists a sequence $t_{n,\delta} \to \infty$ such that

$$\mathbb{P}\left(r_{t_{n,\delta}}(x,y) > \delta\right) \to 0 \text{ as } n \to \infty$$

for $(\rho \times \rho)$-almost every $(x,y) \in M \times M$. A standard "subsequence of subsequence" argument using the monotonicity of $\mathbb{P}\left(r_t(x,y) > \delta\right)$ in $\delta$ now assures the existence of a sequence $t_n \to \infty$ such that for $(\rho \times \rho)$-almost every $(x,y) \in M \times M$ and all $\delta > 0$, $\mathbb{P}\left(r_{t_n}(x,y) > \delta\right) \to 0$ as $n \to \infty$, as required.

(ii) Suppose $\sigma$ is a stationary probability for the two-point motion on $\hat{M}$, with decomposition $\sigma = \sigma_1 + \sigma_2$ where $\sigma_1$ and $\sigma_2$ are respectively absolutely continuous and singular with respect to $(\rho \times \rho)\,|_{\hat{M}}$. Then for all $t > 0$ and $\delta > 0$

$$\int_M \mathbb{P}\left(r_t(x,y) \le \delta\right) d\sigma_1(x,y) \le \sigma\left(\hat{M} \setminus U_\delta\right)$$

Take $t = t_n$, let $n \to \infty$, and use (i) above. We have

$$\sigma_1\left(\hat{M}\right) \le \sigma\left(\hat{M} \setminus U_\delta\right)$$

Now let $\delta \to 0$ to obtain $\sigma_1(\hat{M}) = 0$ as required.

(iii) $\bar{\mu}(\hat{M}) = 0$ implies $(\mu_\omega \times \mu_\omega)(\hat{M}) = 0$ for $\mathbb{P}$-almost every $\omega \in \Omega$ and the result follows immediately. Q.E.D.

**(2.8) Proposition.** *Assume* $\bar{\mu}(\hat{M}) > 0$.

(i) *There exists a stationary probability measure for the two-point motion on* $\hat{M}$.

(ii) *For* $\bar{\mu}$*-almost all* $(x, y) \in \hat{M}$

$$\mathbb{P}\left(r_t(x, y) \to 0 \text{ as } t \to \infty\right) = 0.$$

(iii) *If* $\bar{\mu}(\hat{M}) = 1$ *then* $\mu_\omega$ *is atomless almost surely.*

*Proof.*

(i) Define $\sigma \in \mathcal{P}(\hat{M})$ by $\sigma(A) = \bar{\mu}(A)/\bar{\mu}(\hat{M})$ for $A \in \mathcal{B}(\hat{M})$. The stationarity of $\sigma$ follows from Proposition 2.6(ii) together with the invariance of $\hat{M}$ for the two-point motion.

(ii) Let $\sigma$ be as above. Then $\sigma$ is equivalent to $\bar{\mu}|_{\hat{M}}$, and the validity of (ii) for $\sigma$-almost every $(x, y) \in \hat{M}$ follows from (i) by a general result valid for any stationary Markov process in a non-compact state space.

(iii) Since $\hat{M} = M \times M \setminus \Delta$, the assumption implies $(\mu_\omega \times \mu_\omega)(\Delta) = 0$ for $\mathbb{P}$-almost $\omega \in \Omega$, which in turn implies that $\mu_\omega$ is atomless almost surely. Q.E.D.

## 3. Lyapunov Exponents

For each $x \in M$ and $t \geq 0$ the derivative $D\xi_t(x)$ is a random linear mapping from $T_xM$ to $T_{\xi_t(x)}M$. We investigate the growth rate of $D\xi_t(x)$ as $t \to \infty$. If we impose a smooth Riemannian structure on $M$ we may consider $\left[D\xi_t(x)^* D\xi_t(x)\right]^{1/2} : T_xM \to T_xM$, the positive part of $D\xi_t(x)$ under a polar decomposition. The following theorem is based on the multiplicative ergodic theorem of Oseledec [18]. Its present formulation for stochastic flows of diffeomorphisms is due to Carverhill [5].

**(3.1) Theorem.** *Assume $\rho$ is ergodic for the one-point motion on $M$. For $(\mathbb{P} \times \rho)$-almost every $(\omega, x) \in \Omega \times M$*

$$[D\xi_t(x)^* D\xi_t(x)]^{1/2t} \to \Lambda_{\omega,x} \quad t \to \infty \tag{3.2}$$

*where $\Lambda_{\omega,x}$ is a random linear mapping of $T_xM$ into itself with non-random eigenvalues*

$$e^{\lambda_1} \geq e^{\lambda_2} \geq \ldots \geq e^{\lambda_N} \tag{3.3}$$

*Moreover if $v \in T_xM$ then*

$$\frac{1}{t} \log |D\xi_t(x)(v)| \to \lambda_i \quad \textit{as } t \to \infty$$

$$\textit{if } v \in \bigoplus_{\lambda_j \leq \lambda_i} \ker\left(\Lambda_{\omega,x} - e^{\lambda_j} I\right) \setminus \bigoplus_{\lambda_j < \lambda_i} \ker\left(\Lambda_{\omega,x} - e^{\lambda_j} I\right).$$

The values $\lambda_1 \geq \lambda_2 \geq \ldots \geq \lambda_N$ are called the Lyapunov exponents. They depend on the law of $\{\xi_t : t \geq 0\}$ and $\rho$ but not on the choice of Riemannian structure on $M$. It is clear that if $\lambda_1 < 0$ then for $\rho$-almost every $x \in M$ and all $v \in T_xM$, $|D\xi_t(x)(v)| \to 0$ as $t \to \infty$ almost surely, so that the linearized system is stable. The behavior of $|D\xi_t(x)(v)|$ for the case $\lambda_1 > 0$ is slightly more complicated since it depends on the position of $v$ relative to $N_{\omega,x} \equiv \oplus_{\lambda_j \leq 0} \ker\left(\Lambda_{\omega,x} - e^{\lambda_j} I\right)$. The assumption that $\lambda_1 > 0$ implies that $N_{\omega,x}$ has (fixed) positive codimension in $T_xM$. If there is enough randomness in the stochastic flow $\{\xi_t : t \geq 0\}$ then $N_{\omega,x}$ will be sufficiently randomly positioned inside $T_xM$ that for $\rho$-almost $x \in M$ and all $v \in T_xM \backslash \{0\}$, $\mathbb{P}(v \in N_{\omega,x}) = 0$. In this case $\mathbb{P}(|D\xi_t(x)(u)| \to \infty \text{ as } t \to \infty) = 1$ and the linearized system is unstable. We do not pursue this matter here except to point out the stronger assumptions needed in Section 4 to handle the case $\lambda_1 > 0$ rather than $\lambda_1 < 0$.

Associated with the Lyapunov exponents is the moment Lyapunov function $\tilde{\Lambda} : \mathbb{R} \to \mathbb{R}$ defined by

$$\tilde{\Lambda}(p) = \lim_{t \to \infty} \frac{1}{t} \mathbb{E} |D\xi_t(x)(v)|^p, \quad v \in TM, |v| \neq 0, p \in \mathbb{R}.$$

Under the assumption (4.2) of the next section, $\tilde{\Lambda}$ is well-defined, i.e. the limit exists and is independent of the choice of $v$. Moreover $\tilde{\Lambda}$ is a convex analytic function of $p$ with $\tilde{\Lambda}(0) = 0$ and $\tilde{\Lambda}'(0) = \lambda_1$ (see Baxendale and Stroock [4], equation (2.9) and Theorem 1.29). We shall use properties of $\tilde{\Lambda}$ in Sections 4 and 5.

More information on Lyapunov exponents and the moment Lyapunov function may be obtained from various papers in the collection *Lyapunov Exponents* edited by Arnold and Wihstutz [1].

## 4. Results for $\lambda_1 \neq 0$

Henceforth in this paper we shall assume that $\rho$ is ergodic for the one-point motion on $M$. We list here some other hypotheses which we will assume at various places in Sections 4 and 5.

For $0 \leq k \leq d$, let $TX_k, \tilde{X}_k$, and $GLX_k$ denote the natural lifts of the vector field $X_k$ on $M$ to vector fields on $TM$ ( = tangent bundle of $M$), $SM$ (= unit sphere bundle in $TM$), and $GLM$ ( = linear frame bundle of $M$) respectively. For $v \in SM$ and $u \in C([0,\infty), \mathbb{R}^d)$, let $\Psi(\cdot, v; u)$ denote the curve in $SM$ satisfying

$$\dot{\Psi}(t, v; u) = \sum_{k=1}^{d} u_k(t) \tilde{X}_k(\Psi(t, v; u)) + \tilde{X}_0(\Psi(t, v; u))$$

with $\Psi(0, v; u) = v$.

(4.1) There is no non-empty compact $C \subset \hat{M}$ which is invariant for the two-point motion.

(4.2) $\text{Lie}\left(\tilde{X}_0, \tilde{X}_1, \dots, \tilde{X}_d\right)(v) = T_v SM$ and $\left\{\Psi(t, v; u) : t \geq 0 \text{ and } u \in C\left([0,\infty), R^d\right)\right\}$ is dense in $SM$ for all $v \in SM$

(4.3) $\text{Lie}\ (TX_1, \dots, TX_d)(v) = T_v TM$ for all $v \in TM$ with $|v| \neq 0$.

(4.4) $\text{Lie}\,(GLX_0, GLX_1, \dots, GLX_d)(f) = T_f GLM$ for all $f \in GLM$.

Hypotheses (4.1)–(4.3) are taken from Baxendale and Stroock [4] and (4.4) is taken from Ledrappier and Young [15]. See these papers for more details. Observe that (4.3) implies (4.2) which in turn implies that $L$ is hypoelliptic and that $\rho$ is unique and has smooth positive density with respect to Riemannian measure on $M$. Hypothesis (4.1) implies that the two-point motion from any starting position in $\hat{M}$ will almost surely hit arbitrarity small neighborhoods of the diagonal $\Delta$, while hypotheses (4.2) and (4.3) will be used in the analysis of the two-point motion while it is close to $\Delta$. Each of the four hypotheses concerns the non-degeneracy of a finite-dimensional diffusion process associated with the stochastic flow; the diffusions are on $\hat{M}, SM, TM$ and $GLM$ respectively. The vector fields appearing in (4.2), (4.3) and (4.4) are the vector fields appearing in the stochastic differential equations for the induced diffusions on $SM, TM$ and $GLM$ respectively. (The vector fields for the diffusion on $\hat{M}$ will appear in (6.1).) Thus each hypothesis can be regarded as a particular manifestation of the general hypothesis "assume that there is enough noise driving the stochastic flow". Since the number $d$ of noise-driven vector fields in equation (1.1) is unrelated to the dimension $N$ of the manifold $M$, it is easy to obtain examples where the hypotheses above are satisfied.

The first set of results deals with the case when the top Lyapunov exponent $\lambda_1$ is negative. For $R > 0$ define $\tau_R(x,y) = \inf\{t \geq 0 : r_t(x,y) = R\}$.

**(4.5) Theorem.** *Assume $\lambda_1 < 0$.*

(i) *There exists $n \geq 1$ such that almost surely $\mu_\omega$ consists of $n$ point measures of mass $1/n$ each.*
(ii) *For all $\varepsilon > 0$ there exists $D \in \mathcal{B}(M)$ and $r > 0$ such that $\rho(D) < \varepsilon$ and if $x \notin D$ and dist$(x,y) \leq r$ then $\mathbb{P}(r_t(x,y) \to 0 \text{ as } t \to \infty) \geq 1 - \varepsilon$*

*Proof.* (i) is Proposition 2 of Le Jan [16], while (ii) is an immediate consequence of the local stable manifold for stochastic flows (see Theorem 2.2.1 of Carverhill [5]). Q.E.D.

**(4.6) Theorem.** *Assume $\lambda_1 < 0$ and (4.2). Then we may take $D = \emptyset$ in Theorem 4.5(ii).*

*Proof.* We use the notation and ideas of the proof of Theorem 3.19(v) of Baxendale and Stroock [4] and adapt them to the case $\lambda_1 < 0$. Since $\tilde{\Lambda}(0) = 0$ and $\tilde{\Lambda}(0) = \lambda_1 < 0$ then there exists $\sigma > 0$ such that $\tilde{\Lambda}(\sigma) \leq 0$. The fact that $\varphi_\sigma^- (\xi_t (x), \xi_t (y))$ stopped at time $\tau_\varepsilon (x,y) \wedge \tau_R(x,y)$ is a supermartingale may be used in a calculation similar to (3.23) of [4] to prove the existence of $K < \infty$ and $\delta > 0$ such that

$$\mathbf{P}\left(\tau_\varepsilon(x,y) < \tau_R(x,y)\right) \geq \frac{R^\sigma - K^2\left(dist(x,y)\right)^\sigma}{R^\sigma - \varepsilon^\sigma} \tag{4.7}$$

whenever $0 < \varepsilon < dist(x,y) < R < \delta$. Then

$$\begin{aligned} \mathbf{P}\left(\tau_\varepsilon(x,y) = \infty\right) &\geq \lim_{\varepsilon \to 0} \mathbf{P}\left(\tau_\varepsilon (x,y) < \tau_R(x,y)\right) \\ &\geq \frac{R^\sigma - K^2\left(dist(x,y)\right)^\sigma}{R^\sigma} \end{aligned} \tag{4.8}$$

Moreover for all $R < \delta$

$$\begin{aligned} &\mathbf{P}(r_t(x,y) \to 0 \text{ as } t \to \infty) \\ &\quad \geq \mathbf{P}\left(r_t(x,y) \to 0 \text{ as } t \to \infty \text{ and } \tau_R(x,y) = \infty\right) \\ &\quad = \lim_{\varepsilon \to 0} \mathbf{P}\left(r_t(x,y) < \varepsilon \text{ eventually and } \tau_R(x,y) = \infty\right) \\ &\quad \geq \lim_{\varepsilon \to 0} \limsup_{\gamma \to 0} \mathbf{P}\big(\tau_\gamma(x,y) < \tau_R(x,y) \text{ and } \\ &\qquad\qquad\qquad r_t(x,y) < \varepsilon \text{ for } t > \tau_\gamma(x,y)\big) \\ &\quad \geq \lim_{\varepsilon \to 0} \limsup_{\gamma \to 0} \frac{R^\sigma - K^2\left(dist(x,y)\right)^\sigma}{R^\sigma - \gamma^\sigma} \cdot \frac{\varepsilon^\sigma - K^2\gamma^\sigma}{\varepsilon^\sigma} \\ &\quad = \frac{R^\sigma - K^2\left(dist(x,y)\right)^\sigma}{R^\sigma} \end{aligned} \tag{4.9}$$

where the penultimate line uses (4.7) and (4.8) together with the strong Markov property for the two-point motion at time $\tau_\gamma(x,y)$. Since $K$ and $\delta$ are fixed the right hand side of (4.9) tends to 1 as $dist(x,y) \to 0$. Q.E.D.

**(4.10) Theorem.** *Assume $\lambda_1 < 0$ and (4.1) and (4.2).*

(i) $\mathbf{P}\left(r_t(x,y) \to 0 \text{ as } t \to \infty\right) = 1$ *for all* $(x,y) \in \hat{M}$.
(ii) $\bar{\mu}(\hat{M}) = 0$
(iii) $\mu_\omega$ *consists of a single atom almost surely, and $Q$ is the image of $\rho$ under the map $M \to \mathcal{P}(M)$, $x \mapsto$ unit mass at $x$.*

*Proof.* Let $(x,y) \in \hat{M}$. Given $\varepsilon > 0$ choose $r > 0$ as in Theorem 4.6. Suppose first $dist(x,y) < r$. Then $\mathbf{P}\left(r_t(x,y) \to 0 \text{ as } t \to \infty\right) \geq 1 - \varepsilon$. Now suppose $dist(x,y) \geq r$. By assumption (4.1), and Proposition 4.1 of [4], we have $\tau_{r/2}(x,y) < \infty$ almost surely. Using the strong Markov property at time $\tau_{\sigma/2}(x,y)$ we obtain $\mathbf{P}\left(r_t(x,y) \to 0 \text{ as } t \to \infty\right) \geq 1-\varepsilon$ in this case also. Since $\varepsilon$ was arbitrary we have proved (i). Result (ii) follows from (i) by Proposition 2.8(ii), and the first-part of (iii) follows from (ii) by Proposition 2.7(iii). Finally, the stationary behavior of statistical equilibrium implies that the position of the atom of $\mu_\omega$ has a distribution which is stationary for the one point motion, and assumption (4.2) implies that $\rho$ is the unique stationary probability measure for the one-point motion. Q.E.D.

*(4.11) Remark.* Without assumption (4.1) the case $n > 1$ in Theorem 4.5(i) can occur. For an example with $n = 2$, let $M = S^1 = \mathbf{R}/2\pi\mathbf{Z}$ and consider

$$d\xi_t(\theta) = (\sin 2\xi_t(\theta)) \circ dW_t^1 + (\cos 2\xi_t(\theta)) \circ dW_t^2$$

for $\theta \in S^1$. The resulting stochastic flow is the double cover of the stochastic flow on $S^1$ given as Example 2 in [2]. The stochastic flow does not satisfy (4.2) since it preserves antipodal pairs of points (that is, pairs of points distance $\pi$ apart). In fact the set $C$ consisting of all antipodal pairs provides the contradiction to (4.2). However (4.1) is easily verified, and a calculation similar to that in [2] yields $\lambda_1 = -2$. The statistical equilibium consists almost surely of antipodal pairs of point measures of mass $\frac{1}{2}$ each, and $\bar{\mu}\left(\hat{M}\right) = \bar{\mu}(\Delta) = \frac{1}{2}$.

We now turn to the case when $\lambda_1$ is positive. Since there is no counterpart to the local stable manifold theorem when $\lambda_1 > 0$, our results need at least assumption (4.2).

*(4.12) Remark.* Assume $\lambda_1 > 0$ and (4.2).

(i) There exists $\delta > 0$ such that $\mathbf{P}(\tau_R(x,y) < \infty) = 1$ and $\mathbf{E}(\tau_R(x,y)) < \infty$ whenever $0 < dist(x,y) < R < \delta$.

(ii) $\bar{\mu}\left(\hat{M}\right) = 1$.

(iii) $\mu_\omega$ is atomless almost surely.

*Proof.* Result (i) is contained in Theorem 3.19(i) and (ii) of [4]. To prove (ii) we adapt Theorem III.5.1 of Has'minskii [11]. Theorem 3.18 of [4] assures the existence of a smooth $V\colon \hat{M} \to \mathbf{R}$ satisfying $V \geq 0$, $L^{(2)}V(x,y) \to -\infty$

as $dist(x,y) \to 0$ and $\int_{\hat{M}} V d(p \times \rho) < \infty$. (Here $L^{(2)}$ denotes the generator for the two-point motion; take $V$ to be a smooth non-negative extension of $\varphi_p^-$ for any $p < \infty$ such that $\tilde{\Lambda}(p) < 0$). If we integrate Has'minskii's pointwise estimate with respect to $\rho \times \rho$ we obtain

$$\frac{1}{t}\int_0^t \int_{\hat{M}} \mathbb{P}\left(r_s(x,y) < \delta\right) d\left(\rho \times \rho\right)(x,y) ds \to 0$$

as $\delta \to 0$ uniformly for $t > 0$. In other words for $t > 0$ the probability measures $\frac{1}{t}\int_0^t P_s^*\left(\rho \times \rho\right) ds$ are uniformly tight in $\mathcal{P}(\hat{M})$. Notice here that (4.2) implies $\rho$ is atomless and hence $\rho \times \rho$ and $P_s^*\left(\rho \times \rho\right)$ are all supported on $\hat{M}$. Since $P_t^*\left(\rho \times \rho\right)$ converges weakly to $\bar{\mu}$ in $\mathcal{P}\left(M \times M\right)$, it follows that $\bar{\mu}\big|_{\hat{M}}$ is a probability measure on $\hat{M}$, as required. Result (iii) now follows from (ii) by Proposition 2.8(iii). Q.E.D.

*(4.13) Remark.* Using the terminology of Has'minskii [11], we may restate Theorem 4.12(i) by saying that the two-point motion on $\hat{M}$ is positive recurrent relative to the sets $U_R = \left\{(x,y) \in \hat{M} : dist(x,y) > R\right\}$ for all $R < \delta$.

**(4.14) Theorem.** *Assume $\lambda_1 > 0$ and (4.1) and (4.3).*

(i) *For all $(x,y) \in \hat{M}$ and all bounded measurable $f : \hat{M} \to \mathbb{R}$*

$$\mathbb{P}\left(\frac{1}{t}\int_0^t f\left(\xi_s\left(x\right), \xi_s\left(y\right)\right) ds \to \int_{\hat{M}} f d\bar{\mu} \;\; as \;\; t \to \infty\right) = 1$$

(ii) *$\bar{\mu}$ is the unique stationary probability measure for the two-point motion on $\hat{M}$.*

(iii) *There exist $\gamma \in (0, N]$ and and $K < \infty$ such that*

$$\frac{1}{K} r^{\gamma} \leq \bar{\mu}\left(\hat{M} \setminus U_r\right) \leq K r^{\gamma}$$

*for all sufficiently small $r > 0$.*

*Proof.* This is a restatement of Theorem 4.6 of [4]. Notice that $-\gamma$ is the second zero of the moment Lyapunov function $\tilde{\Lambda}$. Q.E.D.

In the case $\lambda_1 < 0$ with enough noise we obtained a complete characterization of the statistical equilibrium. The situation when $\lambda_1 > 0$ is more complicated. All we have shown so far is that (with enough noise) $\mu_\omega$ is atomless almost surely. The following theorem due to Le Jan ([16], Proposition 1) is relevant here. For convenience we write $\lambda_1 + \ldots + \lambda_N = \lambda_\Sigma$.

**(4.15) Theorem.**

(i) *If $\lambda_\Sigma < 0$ then $\mu_\omega$ is singular with respect to $\rho$ almost surely.*
(ii) *If $\rho$ is equivalent to the Riemannian measure on $M$ then $\lambda_\Sigma \leq 0$. Moreover $\lambda_\Sigma = 0$ if and only if $\rho_t = \rho$ for all $t \geq 0$ almost surely.*

**(4.16) Corollary.** *Assume $\lambda_1 > 0$ and (4.2). Then either*

(a) *$\lambda_\Sigma < 0$ and $\mu_\omega$ is atomless and singular with respect to $\rho$ almost surely; or*
(b) *$\lambda_\Sigma = 0$ and $\mu_\omega = \rho$ almost surely.*

*Proof.* This is immediate from Theorems 4.12 and 4.15 once we have noted that (4.2) implies that $\rho$ has a smooth positive density with respect to Riemannian measure on $M$. Q.E.D.

*(4.17) Remark.* If we assume (4.1) and (4.3) then the following are equivalent

(i) $\lambda_\Sigma = 0$
(ii) $\xi_t$ preserves $\rho$ for all $t \geq 0$ almost surely
(iii) $\mu_\omega = \rho$ almost surely
(iv) $\bar{\mu} = \rho \times \rho$
(v) $\gamma = N$ (where $\gamma$ is the exponent in Theorem 4.14(iii)).

The implication (v)$\Rightarrow$ (i) is contained in Corollary 2.14 of [4], and (i)$\Rightarrow$ (ii)$\Rightarrow$ (iii)$\Rightarrow$ (iv)$\Rightarrow$ (v) are immediate from the results above.

Corollary 4.16 suggests that when $\lambda_1 > 0$ and $\lambda_\Sigma < 0$ the distribution of mass in a typical $\mu_\omega$ is more concentrated than the distribution $\rho$, although not sufficiently concentrated to contain any atoms. This is also suggested by figures 2–4 above. We quantify the degree of clustering in a typical $\mu_\omega$ as follows.

Recall the probability measure $\bar{\mathbb{P}}$ on $\Omega \times M$ with marginal $\mathbb{P}$ and regular conditional probability density $\{\mu_\omega : \omega \in \Omega\}$. Consider the family $\{V_r : r > 0\}$ of random variables defined on $(\Omega \times M, \mathcal{F} \times B(M), \bar{\mathbb{P}})$ by

$$V_r(\omega, x) = \mu_\omega(B(x, r))$$

where $B(x,r) = \{y \in M : dist(x,y) < r\}$. Thus $V_r$ measures how much mass a typical $\mu_\omega$ assigns to the ball of radius $r$ centered at a typical point $x$ chosen with distribution $\mu_\omega$. In the atomless case $V_r \to 0$ as $r \to 0$ almost surely $(\bar{\mathbb{P}})$ and hence also in $L^1(\bar{\mathbb{P}})$. The next result quantifies in two different ways the rate at which $V_r \to 0$.

**(4.18) Theorem.** *Assume* $\lambda_1 > 0$.

(i) *Assume (4.1) and (4.3). Then, for* $K$ *and* $\gamma$ *as in Theorem 4.14(iii) and all sufficiently small* $r$,

$$\frac{1}{K} r^\gamma \le \mathbb{E}^{\bar{\mathbb{P}}}(V_r) \le K r^\gamma$$

(ii) *Assume* $\lambda_i \neq 0$ *for all* $i$ *and (4.4). Then*

$$\lim_{r \to 0} \frac{\log V_r}{\log r} = \sigma \qquad \textit{almost surely } (\bar{\mathbb{P}})$$

*where* $\sigma$ *is the Lyapunov dimension determined by the Lyapunov exponents* $\lambda_1 \geq \ldots \geq \lambda_N$.

*Proof.*

(i) Observe that $\mathbb{E}^{\bar{\mathbb{P}}}(V_r) = \bar{\mu}\left(\hat{M} \setminus U_r\right)$. The result is now a restatement of Theorem 4.14(iii).

(ii) This is Theorem $A'$ of Ledrappier and Young [15]. Their paper contains some alternative hypotheses under which the result is also valid.

Q.E.D.

*(4.19) Remark.* The result of Ledrappier and Young above implies that under the conditions of Theorem 4.18(ii), $\mu_\omega$ almost surely has Hausdorff dimension $\sigma$, in the sense that

a) there exists $F \in \mathcal{B}(M)$, depending on $\mu_\omega$, such that Haus-dim$(F) = \sigma$ and $\mu_\omega(F) = 1$; and
b) if Haus-dim$(G) < \sigma$ for $G \in \mathcal{B}(M)$ then $\mu_\omega(G) = 0$.

*(4.20) Remark.* A comparison of (i) and (ii) in Theorem 4.18 shows that $\gamma \leq \sigma$. It can happen that $\gamma < \sigma$. For example in the stochastic flow simulated in Section 2 we have $\gamma \sim 1.17$ and $\sigma \sim 1.59$. [These are approximate values obtained by numerical integration; the author wishes to thank Dr. John Pulham of University of Aberdeen, Scotland, for assistance in obtaining these values.]

## 5. Results for $\lambda_1 = 0$

Sections 3 and 4 of Baxendale and Stroock [4] study the behavior of the two-point motion on $\hat{M}$ in the case when $\lambda_1 > 0$. Here we will use similar methods to study the remaining case $\lambda_1 = 0$. Recall that $\tilde{\Lambda}$ is the moment Lyapunov function for the stochastic flow, and $\tilde{\Lambda}'(0) = \lambda_1$.

**(5.1) Proposition.** *Assume (4.2). If $\lambda_1 = 0$ and $\tilde{\Lambda}''(0) = 0$ then there exists a smooth Riemannian structure on $M$ with respect to which $\xi_t$ is an isometry for all $t \geq 0$ almost surely.*

*Proof.* We use the notation and results contained on pages 198–99 of [4]. For the moment let $\lambda_1$ take any value. Differentiate equation (3.15) of [4] with respect to $p$ and put $p = 0$ to obtain

$$\tilde{L}\varphi_0'' + 2\left(\sum_{k=1}^{d} g_k \tilde{X}_k + \tilde{Q} - \lambda_1\right)\varphi_0' + \tilde{a} = \tilde{\Lambda}''(0)\,.$$

Combining this with (3.15) evaluated at $p = 0$ leads to

$$\tilde{L}\left(\varphi_0'' - (\varphi_0')^2\right) + \sum_{k=1}^{d}\left(\tilde{X}_k\varphi_0' + g_k\right)^2 = \tilde{\Lambda}''(0)\,.$$

Now integrate over $SM$ with respect to $\tilde{m}_{\tilde{L}}$. If $\tilde{\Lambda}''(0) = 0$ then

$$\tilde{X}_k \varphi_0' = -g_k \qquad \text{for } 1 \leq k \leq r.$$

Since $\tilde{L}\varphi_0' = \lambda_1 - \tilde{Q}$ then

$$\tilde{X}_0 \varphi_0' = \lambda_1 - g_0$$

Therefore, writing $r_t = |D\xi_t(x)(v)|$ and $\theta_t = \frac{D\xi_t(x)(v)}{|D\xi_t(x)(v)|} \in SM$ we have, by Itô's formula,

$$d(\log r_t + \varphi_0'(\theta_t)) = \lambda_1 dt.$$

Hence

$$r_t = r_0 \exp(\varphi_0'(\theta_t) - \varphi_0'(\theta_t) + \lambda_1 t)$$

which implies, by the boundedness of $\varphi_0'$, that $\tilde{\Lambda}(p) = \lambda_1 p$ for all $p \in \mathbf{R}$. Now suppose also $\lambda_1 = 0$. Then $\tilde{\Lambda} \equiv 0$ and the result follows by Corollary 2.14 of [4]. Q.E.D.

The result above, together with the convexity of $\tilde{\Lambda}$, implies that if we assume $\lambda_1 = 0$ then we may also assume, with hardly any loss of generality, that $\tilde{\Lambda}''(0) > 0$. In particular the assumptions (4.3) and $\lambda_1 = 0$ together imply $\tilde{\Lambda}''(0) > 0$ since if $\xi_t$ consists almost surely of isometries for all $t \geq 0$ then the vector fields $X_0, X_1, \ldots, X_d$ must all be infinitesimal isometries, contradicting (4.3). We shall write $\tilde{\Lambda}''(0) = V$ (since $\tilde{\Lambda}''(0)$ is the variance in the central limit theorem for $\log|D\xi_t(x)(v)|$, see Theorem 1.29 (iv) of [4]). In the following, $L^{(2)}$ is the generator of the two-point motion and $(r, \varphi) \in (0, \delta) \times SM$ denote the polar coordinates of a point $(x, y)$ satisfying $0 < dist(x, y) < \delta$.

**(5.2) Proposition.** *Assume (4.2), $\lambda_1 = 0$ and $V > 0$. Then there exist $\delta > 0$ and $K < \infty$ and functions $\psi^+, \psi^-, \eta^+, \eta^- \in C^\infty((0, \delta) \times SM)$ such that*

$$L^{(2)}\psi^+ \geq 0 \geq L^{(2)}\psi^- \quad \text{on } (0, \delta) \times \mathrm{S}M$$

$$L^{(2)}\eta^+ \geq V \geq L^{(2)}\eta^- \quad \text{on } (0, \delta) \times \mathrm{S}M$$

$$|\psi^\pm(r, \theta) - \log r| \leq K \qquad (r, \theta) \in (0, \delta) \times \mathrm{S}M$$

$$\left|\eta^\pm(r, \theta) - (\log r)^2\right| \leq K|\log r| \qquad (r, \theta) \in (0, \delta) \times \mathrm{S}M$$

*Proof.* The functions $\psi^+$ and $\psi^-$ are given by Theorem 3.18(ii) of [4]. The existence of suitable $\eta^+$ and $\eta^-$ is proved by a similar method. We give some details. Notice first that

$$\mathbf{T}L\left((\log r)^2 + 2(\log r)\varphi_0'(\theta) + \varphi_0''(\theta)\right) = V$$

Remark (3.11) of [4] can be strengthened by observing that whenever $\psi \equiv 1$ the estimates in (3.9) and (3.10) of [4] remain valid when the $r|\phi(r)|$ terms on the right hand sides are omitted. Therefore there exist $\delta > 0$ and $K < \infty$ such that

$$\left|\left(TL - L^{(2)}\right)\left((\log r)^2 + 2(\log r)\varphi_0'(\theta) + \varphi_0''(\theta)\right)\right| \leq Kr\,|\log r|$$

for $(r,\theta) \in (0,\delta) \times SM$. The assumptions $\lambda_1 = 0$ and $V > 0$ imply $\tilde{\Lambda}\left(\frac{1}{2}\right) > 0$. Take

$$\eta^{\pm}(r,\theta) = (\log r)^2 + 2(\log r)\varphi_0'(\theta) + \varphi_0''(\theta) \pm kr^{\frac{1}{2}}\varphi_{\frac{1}{2}}(\theta)$$

for suitable $k$ (to be chosen later). The remainder of the proof is similar to that given for Theorem 3.18 in [4]. Q.E.D.

The existence of the functions $\psi^{\pm}$ and $\eta^{\pm}$ implies that, when $r_t(x,y)$ is small, $\log r_t(x,y)$ behaves, modulo lower order terms, like a martingale with quadratic variation growing at rate $V = \tilde{\Lambda}''(0)$. We make this idea precise in the following result.

**(5.3) Theorem.** *Assume (4.2), $\lambda_1 = 0$ and $V > 0$. There exist $\delta > 0$ and $K < \infty$ such that if $0 < \varepsilon < dist(x,y) < R < \delta$ then*

(i) $$\mathbf{P}\left(\tau_\varepsilon(x,y) \wedge \tau_R(x,y) < \infty\right) = 1$$

(ii) $$\frac{\log\left(\dfrac{R}{dist(x,y)}\right) - 2K}{\log\left(\dfrac{R}{\varepsilon}\right)} \leq \mathbf{P}\left(\tau_\varepsilon(x,y) < \tau_R(x,y)\right) \leq \frac{\log\left(\dfrac{R}{dist(x,y)}\right) + 2K}{\log\left(\dfrac{R}{\varepsilon}\right)}$$

$$\text{(iii)}\quad \begin{aligned} &\frac{1}{V}\left[\log\left(\frac{R}{dist(x,y)}\right)\log\left(\frac{dist(x,y)}{\varepsilon}\right) - 6K\,|\log\varepsilon| + 2K^2\right] \\ &\le \mathsf{E}\left(\tau_\varepsilon(x,y)\wedge\tau_R(x,y)\right) \\ &\le \frac{1}{V}\left[\log\left(\frac{R}{dist(x,y)}\right)\log\left(\frac{dist(x,y)}{\varepsilon}\right) + 6K\,|\log\varepsilon| + 2K^2\right] \end{aligned}$$

*Proof.* The method of proof is essentially the same as that used for Theorem (3.19) in [4]. Let the process $\{(r_t(x,y),\theta_t(x,y)) : t \ge 0\}$ in $(0,\delta] \times SM$ be the polar coordinate expression for the two-point motion

$$\{(\xi_t(x),\xi_t(y)) : t \ge 0\}$$

stopped at time $\tau_\delta(x,y)$. (This agrees with our previous usage of $r_t(x,y)$; do not confuse with $\{(r_t,\theta_t) : t \ge 0\}$ in the proof of Proposition 5.1 which corresponded to the linearized process.) Then (i) follows from the fact that $\eta^+\left(r_t(x,y),\theta_t(x,y)\right) - Vt$, stopped at $\tau_\varepsilon(x,y)\wedge\tau_R(x,y)$, is a submartingale. The proof of (ii) follows that of Theorem 3.19(v), using $\psi^\pm$ in place of $\varphi^\pm_{-\gamma}$. The proof of (iii) uses (ii) together with the fact that

$$\pm\left(\eta^\pm\left(r_t(x,y),\theta_t(x,y)\right) - Vt\right),$$

stopped at $\tau_\varepsilon(x,y)\wedge\tau_R(x,y)$, are both submartingales. Q.E.D.

**(5.4) Corollary.** *Under the conditions of Theorem (5.3)*

(i) $$\mathsf{P}(\tau_R(x,y) < \infty) = 1$$

(ii) $$\mathsf{E}(\tau_R(x,y)) = \infty \ \ \textit{whenever}\ \ dist(x,y) < e^{-6K}R$$

*Proof.* Let $\varepsilon \to 0$ in the above. Q.E.D.

*(5.5) Remark.* The preceding corollary says that the two-point motion in $\hat{M}$ is recurrent but not positive recurrent relative to the sets $U_R$ for all $R < \delta$.

While the corollary above already assures the existence of an (infinite) stationary measure for the two-point motion on $\hat{M}$, the following result will yield an estimate on the rate at which the measure of $U_r$ grows as $r \to 0$.

**(5.6) Proposition.** *Assume (4.2), $\lambda_1 = 0$ and $V > 0$. Let $\delta$ and $K$ be as in Theorem 5.3. For fixed $0 < R < \delta$ define*

$$g_\varepsilon(x,y) = \mathbf{E}\left[\int_0^{\tau_R(x,y)} \chi_{[\varepsilon,\infty)}\left(r_s(x,y)\right) ds\right]$$

*for $0 < \varepsilon \leq dist(x,y) < R$. Then*

$$\frac{2}{V}\left[\log\left(\frac{R}{dist(x,y)}\right) - 4K\right] \leq \liminf_{\varepsilon\to 0} \frac{g_\varepsilon(x,y)}{|\log\varepsilon|}$$
$$\leq \limsup_{\varepsilon\to 0} \frac{g_\varepsilon(x,y)}{|\log\varepsilon|} \leq \frac{2}{V}\left[\log\left(\frac{R}{dist\,(x,y)}\right) + 4K\right].$$

*Proof.* For $\varepsilon$ temporarily fixed and $\varepsilon \leq r < R$, define

$$g^{\pm}(r) = \pm\sup\left\{\pm g_\varepsilon(x,y) : dist(x,y) = r\right\}.$$

By conditioning at time $\tau_\varepsilon(x,y) \wedge \tau_R(x,y)$ we obtain

$$\begin{aligned} &g^{-}(\varepsilon)\mathbf{P}(\tau_\varepsilon(x,y) < \tau_R(x,y)) + \mathbf{E}\left(\tau_\varepsilon(x,y) \wedge \tau_R(x,y)\right) \\ &\quad\leq g_\varepsilon(x,y) \\ &\quad\leq g^{+}(\varepsilon)\,\mathbf{P}\left(\tau_\varepsilon(x,y) < \tau_R(x,y)\right) + \mathbf{E}\left(\tau_\varepsilon(x,y) \wedge \tau_R(x,y)\right). \end{aligned} \tag{5.7}$$

For $\varepsilon < a < R$ the monoticity of $g^-$ together with (5.7) imply

$$\begin{aligned} g^{-}(\varepsilon) &\geq g^{-}(a) \\ &\geq \inf_{dist(u,v)=a}\left[g^{-}(\varepsilon)\,\mathbf{P}\left(\tau_\varepsilon(u,v) < \tau_R(u,v)\right) + \mathbf{E}\left(\tau_\varepsilon(u,v) \wedge \tau_R(u,v)\right)\right] \end{aligned}$$

This yields, using Theorem 5.3(ii) and (iii), a lower bound for $g^-(\varepsilon)$. For $0 < b < \varepsilon$, conditioning at time $\tau_b(u,v) \wedge \tau_R(u,v)$ and then again at the first time thereafter that $r_t(u,v) = \varepsilon$ implies

$$g^{+}(\varepsilon) \leq \sup_{dist(u,v)=\varepsilon}\left[\mathbf{E}\left(\tau_b(u,v) \wedge \tau_R(u,v)\right) + \mathbf{P}\left(\tau_b(u,v) < \tau_R(u,v)\right) g^{+}(\varepsilon)\right].$$

This yields, again using Theorem 5.3(ii) and (iii), an upper bound for $g^+(\varepsilon)$. Substituting these bounds for $g^\pm(\varepsilon)$ into (5.7) yields upper and lower bounds for $g_\varepsilon(x,y)$. The remainder of the proof is an exercise in calculus: choose $a$ and $b$ so as to optimise the bounds on $g_\varepsilon(x,y)$ and then let $\varepsilon \to 0$. Q.E.D.

**(5.8) Theorem.** *Assume (4.1), (4.3) and $\lambda_1 = 0$. Then there exists a stationary $\sigma$-finite measure $\nu$ for the two-point motion on $\hat{M}$, unique up to a multiplicative constant. For all $(x,y) \in \hat{M}$ and all bounded measurable $\nu$-integrable $f, g : \hat{M} \to \mathbb{R}$ with $\int_{\hat{M}} g d\nu \neq 0$*

$$\mathbb{P}\left( \frac{\int_0^t f(\xi_s(x), \xi_s(y))\, ds}{\int_0^t g(\xi_s(x), \xi_s(y))\, ds} \to \frac{\int f d\nu}{\int g d\nu} \text{ as } t \to \infty \right) = 1 \tag{5.9}$$

*Moreover there exists $a \in (0,\infty)$ such that*

$$\nu(U_\varepsilon) / |\log \varepsilon| \to a \text{ as } \varepsilon \to 0.$$

*Proof.* As in Theorem 4.6 of [4] we follow the construction of $\nu$ as given by Has'minskii [10]. Fixing $0 < r < R < \delta$ with $r < e^{-6K} R$ (and $\delta$ as in Theorem 5.3), define times inductively

$$\begin{aligned} \sigma_0(x,y) &= \inf\{t \geq 0 : r_t(x,y) = r\} \\ \sigma_n'(x,y) &= \inf\{t \geq \sigma_n(x,y) : r_t(x,y) = R\} \\ \sigma_{n+1}(x,y) &= \inf\{t \geq \sigma_n'(x,y) : r_t(x,y) = r\}. \end{aligned}$$

Using Proposition 4.1 of [4] and Corollary 5.4 we obtain $\sigma_n'(x,y) < \infty$ for all $n \geq 0$ almost surely with

$$\mathbb{E}\left(\sigma_{n+1}(x,y) - \sigma_n'(x,y)\right) < \infty$$

and

$$\mathbb{E}\left(\sigma_n'(x,y) - \sigma_n(x,y)\right) = \infty.$$

Let $\{Z_n : n \geq 0\}$ be the Markov chain on $\Gamma \equiv \left\{(x, y) \in \hat{M} : dist(x, y) = r\right\}$ given by the positions of the two-point motion at times $\sigma_n(x, y)$; we denote its transition function $\Pi((x, y), \cdot)$. Assumption (4.3) implies that $\{Z_n : n \geq 0\}$ has a unique stationary probability measure $\tilde{\nu}$, say, which is equivalent to $\Pi((x, y), \cdot)$ for each $(x, y) \in \Gamma$. Define, for $A \in \mathcal{B}(\hat{M})$,

$$\nu(A) = \int_\Gamma \mathbb{E}\left[\int_0^{\sigma_1(x,y)} \chi_A(\xi_s(x), \xi_s(y))\, ds\right] d\tilde{\nu}(x, y).$$

The fact that $\nu(A)$ is stationary and the ergodic property (5.9) now follow just as in [10]. The uniqueness of $\nu$ up to a multiplicative constant follows just as in Theorem 4.2 of Maruyama and Tanaka [17]. Finally, since $\mathbb{E}(\sigma_1(x, y) - \sigma_0'(x, y))$ is bounded for $(x, y) \in \Gamma$, then

$$\nu(U_\varepsilon) \sim \int_\Gamma g_\varepsilon(x, y) d\tilde{\nu}(x, y) \quad \text{as } \varepsilon \to 0.$$

Hence Proposition 5.6 implies

$$\begin{aligned} \frac{2}{V}\left[\log\left(\frac{R}{r}\right) - 4K\right] \leq \quad & \liminf_{\varepsilon \to 0} \quad \frac{\nu(U_\varepsilon)}{|\log \varepsilon|} \\ & \leq \limsup_{\varepsilon \to 0} \quad \frac{\nu(U_\varepsilon)}{|\log \varepsilon|} \leq \frac{2}{V}\left[\log\left(\frac{R}{r}\right) + 4K\right]. \end{aligned}$$

In particular $\liminf\limits_{\varepsilon \to 0} = a$, say, with $0 < a < \infty$ and

$$\frac{\limsup\limits_{\varepsilon \to 0} \nu(U_\varepsilon) / |\log \varepsilon|}{\liminf\limits_{\varepsilon \to 0} \nu(U_\varepsilon) / |\log \varepsilon|} \leq \frac{\log\left(\frac{R}{r}\right) + 4K}{\log\left(\frac{R}{r}\right) - 4K} \tag{5.10}$$

Consider a sequence $\{r_n\}$ of positive real numbers converging to 0. Replacing $r$ by $r_n$ in the construction of $\nu$ above yields a sequence $\{\nu_n\}$ of stationary measures. The uniqueness of $\nu$ up to multiples implies that the left side of

(5.10) is unchanged when $\nu$ is replaced by $\nu_n$, whereas replacing $r$ by $r_n$ causes the right side to converge to 1 as $n \to \infty$. Thus

$$\limsup_{\varepsilon \to 0} \nu (U_\varepsilon) / |\log \varepsilon| = \liminf_{\varepsilon \to 0} \upsilon (U_\varepsilon) / |\log \varepsilon| = a$$

and we are done. Q.E.D.

*(5.11) Remark.* Has'minskii's construction is a version for diffusion processes of the method used earlier by Harris [8] to construct stationary measures for certain discrete time Markov processes. In particular $\{Z_n : n \geq 0\}$ is analogous to the "process on $A$" of Harris.

**(5.12) Corollary.** *Assume (4.1), (4.3) and* $\lambda_1 = 0$.

(i) *There is no stationary probability measure for the two-point motion on* $\hat{M}$
(ii) $\bar{\mu}(\hat{M}) = 0$
(iii) $\mu_\omega$ *consists of a single atom almost surely, and* $Q$ *is the image of* $\rho$ *under the map* $M \to \mathcal{P}(M), x \mapsto$ *unit mass at* $x$.

*Proof.* The results follow, in sequence, from Theorem 5.8, Proposition 2.8(i) and Proposition 2.7(ii). Q.E.D.

## 6. Transience and recurrence of the two-point motion

We have already seen in Theorem 4.12, Corollary 5.4, and Theorem 4.10 results on the positive recurrent, null recurrent, or transient behavior of the two-point motion near the diagonal according as $\lambda_1$ is positive, zero, or negative. The following theorem assumes sufficient non-degeneracy of the two-point motion on $\hat{M}$ so as to obtain similar results for the two-point motion on the whole of $\hat{M}$.

For $0 \leq k \leq d$ let $X_k^{(2)}$ denote the vector field on $M \times M$ given by $X_k^{(2)}(x, y) = (X_k(x), X_k(y)) \in T_{(x,y)}(M \times M)$. Consider the following hypothesis

$$\text{(6.1)} \qquad \text{Lie}\left(X_1^{(2)}, \ldots, X_d^{(2)}\right)(x, y) = T_{(x,y)}(M \times M) \ \text{ for all } \ (x, y) \in \hat{M}$$

Notice that the Lie algebra condition in (6.1) will automatically fail when $x = y$ since all the vector fields $X_k^{(2)}$ are tangential to $\Delta$. However it is easy to obtain examples where (6.1) is satisfied. In fact it is satisfied on an open dense subset of $d$-tuples of smooth vector fields as soon as $d \geq 2N$. The support theorem of Stroock and Varadhan [20] implies that (6.1) is stronger then (4.1). The only place in this paper where the additional strength of (4.3) over (4.2) has been used is to show that the Lie algebra condition of (6.1) is valid for $(x, y)$ in some deleted neighborhood of $\Delta$ (see Lemma 4.4 of [4]). Hence all our results which are valid under (4.1) and (4.3) remain valid under (4.2) and (6.1).

For open $U \subset \hat{M}$ let $\bar{U}$ denote the closure of $U$ in $\hat{M}$. We write

$$\tau_U(x, y) = \inf\left\{t \geq 0 : (\xi_t(x), \xi_t(y)) \in \bar{U} \setminus U\right\}.$$

This agrees with our previous notation $\tau_R(x, y)$ when $U = U_R$. We rely on the context to avoid confusion in other cases.

**(6.2) Theorem.** *Assume (4.2) and (6.1).*

(i) *If* $\lambda_1 > 0$ *then for all open* $U \subset \hat{M}$ *and all* $(x, y) \in \hat{M} \setminus U$, $\mathsf{P}(\tau_U(x, y) < \infty) = 1$ *and* $\mathsf{E}(\tau_U(x, y)) < \infty$.

(ii) *If* $\lambda_1 = 0$ *then for all open* $U \subset \hat{M}$ *and all* $(x, y) \in \hat{M} \setminus U$, $\mathsf{P}(\tau_U(x, y) < \infty) = 1$. *Further if* $\bar{U}$ *is compact and if* $(x, y)$ *is in the unbounded component of* $\hat{M} \setminus \bar{U}$ *then* $\mathsf{E}(\tau_U(x, y)) = \infty$.

(iii) *If* $\lambda_1 < 0$ *then for all open* $U$ *with* $\bar{U}$ *compact,* $\mathsf{P}(\tau_U(x, y) < \infty) \to 0$ *as* $dist(x, y) \to 0$, *and* $\mathsf{P}(\tau_U(x, y) < \infty) < 1$ *as soon as* $(x, y)$ *is in the unbounded component of* $\hat{M} \setminus \bar{U}$.

*Proof.*

(i) The assertion is valid for $U = U_R$ for all sufficiently small $R$, by Theorem 4.12(i). The assumption (6.1) ensures sufficient non-degeneracy of the two-point motion that the method of Has'minskiii ([11], Lemmas IV.2.1 and IV.3.1) applies here to give the full result.

(ii) The first part follows as above, using Corollary 5.4. To prove the second part we may assume $U \subset U_R$ for some $R < \delta$ (with $\delta$ as in Theorem 5.3). Choose $r < R$ so that $\mathsf{E}(\tau_R(x, y)) = \infty$ whenever $dist(x, y) \leq r$. For each $(x, y)$ in the unbounded component of $\hat{M} \setminus \bar{U}$

there exists a continuous path $p : [0,1] \to \hat{M}$ such that $p(0) = (x,y)$ and $p(1) \in \hat{M} \setminus \bar{U}_r$ and $p([0,1]) \cap \bar{U} = \emptyset$. (For open subsets of $\hat{M}$, connected implies path-connected.) Moreover the set of paths $p$ satisfying these conditions forms an open set in the space of continuous paths $[0,1] \to \hat{M}$ started at $(x,y)$. It follows from (6.1) and the support theorem of Stroock and Varadhan applied to $\{(\xi_t(x), \xi_t(y)) : t \geq 0\}$ that $\mathbb{P}(\Omega_1) > 0$ for the event

$$\Omega_1 = \{\tau_U(x,y) > 1 \text{ and } dist(\xi_1(x), \xi_1(y)) < r\}$$

On the set $\Omega_1$ we have

$$\begin{aligned}\tau_U(x,y) &\geq \inf\{t \geq 1 : (\xi_t(x), \xi_t(y)) \in \partial U\} \\ &\geq \inf\{t \geq 1 : r_t(x,y) = R\}\end{aligned}$$

By conditioning at time 1 we obtain

$$\begin{aligned}\mathbb{E}(\tau_U(x,y)) &\geq \int_{\Omega_1} \tau_U(x,y) d\mathbb{P} \\ &\geq \mathbb{P}(\Omega_1)(1 + \inf\{\mathbb{E}(\tau_U(u,v)) : dist(u,v) < r\}) \\ &= \infty.\end{aligned}$$

(iii) If $\bar{U}$ is compact then $U \subset U_R$ for some $R < \delta$. For $dist(x,y) < R$

$$\mathbb{P}(\tau_U(x,y) < \infty) \geq \mathbb{P}(\tau_R(x,y) < \infty)$$

and the first result follows from (4.8). The second assertion is proved similarly to (ii) above; in this case we obtain

$$\begin{aligned}&\mathbb{P}(\tau_U(x,y) = \infty) \\ &\qquad \geq \mathbb{P}(\Omega_1) \inf\{\mathbb{P}(\tau_R(u,v) = \infty) : dist(u,v) < r\} > 0\end{aligned}$$

for $r$ sufficiently small relative to $R$. Q.E.D.

*(6.3) Remark.* Using recurrence and transience in the sense of hitting open sets from arbitrary starting points (see for example Has'minskii [11]), the theorem above gives a complete classification of the two-point motion on $\hat{M}$ as positive recurrent, null recurrent, or transient according as $\lambda_1$ is positive, zero, or negative. The proof of this result has been based on the facts that $\lambda_1$ describes the almost sure exponential growth rate of the linearized process $\{D\xi_t(x)(v) : t \geq 0\}$ in $TM$, and that $\{D\xi_t(x)(v) : t \geq 0\}$ may be regarded as the linearization of the two-point motion on $\hat{M}$ near $\Delta$.

## References


[1] L. Arnold and V. Wihstutz (eds), "Lyapunov exponents," Lect. Notes Math., **1186**, Springer, Berlin Heidelberg New York, 1986.

[2] P.H. Baxendale, *Asymptotic behavior of stochastic flows of diffeomorphisms*, in "Stochastic processes and their applications, K. Itô and T. Hida (eds)," Lect. Notes Math. **1203**, Springer, Berlin Heidelberg New York, 1986, pp. 1–19.

[3] P.H. Baxendale and T.E. Harris, *Isotropic stochastic flows*, Ann. Probab. **14** (1986), pp. 1155–1179.

[4] P.H. Baxendale and D.W. Stroock, *Large deviations and stochastic flows of diffeomorphisms*, Probab Th. Rel Fields **80** (1988), pp. 169–215.

[5] A. Carverhill, *Flows of stochastic dynamical systems: ergodic theory*, Stochastics **14** (1985), pp. 273–317.

[6] R.W.R. Darling, *Constructing nonhomeomorphic stochastic flows*, Mem. Amer. Math. Soc. **70, No 376** (1987).

[7] K.D. Elworthy, "Stochastic Differential Equations on Manifolds," L.M.S. Lecture Notes 70, Cambridge University Press, Cambridge, 1982.

[8] T.E. Harris, *The existence of stationary measures for certain Markov processes*, in "Proc. $3^{rd}$ Berkeley Symp. Math. Statist. Prob., II," Univ. Calif. Press, Berkeley, 1956, pp. 113–124.

[9] T.E. Harris, *Brownian motions on the homeomorphisms of the plane*, Ann. Probab. **9** (1981), pp. 232–254.

[10] R.Z. Has'minskii, *Ergodic properties of recurrent diffusion processes and stabilization of the solution to the Cauchy problem for parabolic equations*, Theory Prob. and Appl. **5** (1960), pp. 179–196.

[11] R.Z. Has'minskii, "Stochastic stability of differential equations," Sijthoff and Noordhoff, Alphen aan den Rijn, Netherlands, 1980.

[12] N. Ikeda and S. Watanabe, "Stochastic differential equations and diffusion processes," North Holland-Kodansha, Amsterdam Tokyo, 1981.

[13] Y. Kifer, "Ergodic theory of random transformations," Birkhauser, Basel, 1986.

[14] H. Kunita, *Stochastic differential equations and stochastic flow of diffeomorphisms*, in "École d'Été de Probabilités de Saint-Flour XII," Lect. Notes Math. **1097**, Springer, Berlin Heidelberg New York, 1984.

[15] F. Ledrappier and L.-S. Young, *Dimension formula for random transformations*, Commun. Math. Phys. **117** (1988), pp. 529–548.

[16] Y. Le Jan, *Équilibre statistique pour les produits de difféomorphismes aléatoires indépendants*, Ann. Inst. Henri Poincaré, Sect. B **23** (1987), pp. 111–120.

[17] G. Maruyama and H. Tanaka, *Some properties of one-dimensional diffusion processes*, Mem. Fac. Sci. Kyushu Univ. A-11 (1957), pp. 117–141.

[18] V.I. Oseledec, *A multiplicative ergodic theorem. Lyapunov characteristic numbers for dynamical systems*, Trans. Moscow Math. Soc. **19** (1968), pp. 197–231.

[19] L.C.G. Rogers and D. Williams, "Diffusions, Markov processes and martingales, vol 2: Itô calculus," Wiley, Chichester, 1987.

[20] D.W. Stroock and S.R.S. Varadhan, *On the support of diffusion processes with applications to the strong maximum principle*, in "Proc. $6^{th}$ Berkeley Symp. Math. Statis. Prob., III," Univ. Calif. Press, Berkeley, 1972, pp. 333–359.

[21] L.-S. Young, *Stochastic stability of hyperbolic attractors*, Ergod. Th. and Dynam. Sys. **6** (1986), pp. 311–319.

Department of Mathematics
University of Southern California
Los Angeles, California 90089-1113

# Asymptotic Properties of Isotropic Brownian Flows

Yves le Jan

## 1. Isotropic Brownian Flows

Denote by Diff($M$) the group of $C^\infty$ diffeomorphisms of a Riemannian manifold $M$. Define a *Brownian flow* on $M$ as a Diff($M$)-valued process $\Phi_t$, with *independent increments* ($\Phi_{t_1}, \Phi_{t_2} \circ \Phi_{t_1}^{-1}, \ldots, \Phi_{t_{1+1}} \circ \Phi_{t_i}^{-1}$ if, $t_1 < t_2 \ldots < t_i$). We assume in addition that the increments of the flow are *stationary* $\left(\Phi_s \underset{(law)}{\sim} \Phi_{t+s} \circ \Phi_t^{-1}\right)$ and that for any initial value $x \in M$, the one point process $\Phi_t(x)$ is a Brownian motion on $M$. An analytical characterization of Brownian flows is given in [1].

*Example 1.* On $SL_d(\mathbb{R})$, a left multiplicative Brownian motion $Y_t$ is defined by a Stratonovitch equation:

$$Y_0 = I, \quad d\,Y_j^i(t) = Y_j^k(t) \circ dW_k^i(t)$$

where $W_k^i(t)$ is an additive Brownian motion on the Lie algebra (matrices with zero trace). The choice of $W$ is arbitrary and equivalent to the choice of a left invariant metric. $\Phi_t(X) \underset{(def)}{=} Y_t X$ is then a Brownian flow.

For later use, we can choose $W$ such that:

$$E\left(W_j^i(t)\, W_\ell^k(s)\right) = \alpha\,(t \wedge s)\left[(d+1)\,\delta_k^i \delta_\ell^j - \delta_j^i \delta_\ell^k - \delta_\ell^i \delta_j^k\right]$$

(cf. [9] p. 610). This covariance is isotropic, i.e. $RW \underset{(law)}{\sim} W$ for any rotation $R$. Moreover, Itô corrections vanish in the expression of $dY_t : dY_j^i(t) = Y_j^k(t) dW_k^i(t)$.

For any $x \in R^d$, set $\Pi(x) = \frac{x}{\|x\|}$. Then $\Pi Y_t$ defines a Brownian flow on $\mathbb{S}^{d-1}$.

*Example 2. Stationary isotropic measure preserving Brownian flows on* $\mathbb{R}^d$. To each measure $F(d\rho)$ on $[0,+\infty)$ with finite moments $\mu_n$ of all orders, one can associate an isotropic stationary smooth gaussian vector field on $\mathbb{R}^d$ with zero divergence and zero mean. It is determined by its covariance function:

$$E\left(W^i(x)\, W^j(y)\right) = C^{ij}\,(x-y)$$

with

$$c_{ij}(z) = \int\limits_{\mathbb{R}+} \int\limits_{\mathbb{S}^{d-1}} e^{i\rho\langle z,u\rangle} \left(\delta^i_j - u^i u^j\right) \sigma(du) F(d\rho),$$

$\sigma(du) =$ normalized lebesgue measure on $\mathbb{S}^{d-1}$. (cf [6], [13]).

We can also construct the associated vector field-valued Brownian motion $W^i_t(x)$ determined by its covariance function:

$$E\left(W^i_t(x)\, W^j_s(y)\right) = t \wedge s\;\; C^{ij}\,(x-y)\,.$$

Zero divergence smooth vector fields can be viewed as the Lie algebra associated with the group of measure preserving diffeomorphisms of $\mathbb{R}^d$. By analogy with the preceding example, a Brownian flow $\Phi_t$ can be constructed by solving a generalized stochastic differential equation:

$$\Phi^i_t(x) = x^i + \int_0^t W^i\left(\Phi_s(x), ds\right)$$

(cf. [8],[9],[2],[10],[5]). $\Phi_t$ is stationary: $T_a \Phi_t T_{-a} \underset{(law)}{\sim} \Phi_t$ for any translation $T_a$; and isotropic: $R\Phi_t R^{-1} \sim \Phi_t$ for any rotation $R$. It is characterized by the following property: For any $(x_1, x_2 \ldots x_n) \in \left(\mathbb{R}^d\right)^n, \left(\Phi_t\left(x_1\right), \ldots, \Phi_t\left(x_n\right)\right)$ is a diffusion with generator $\frac{1}{2}\Sigma c^{ij}\left(x_k - x_\ell\right) \frac{\partial}{\partial x^i_k} \frac{\partial}{\partial x^j_\ell}$. We shall now restrict our study to this example. In two dimensions, it was first constructed by T. Harris, as a limit in law of a family of discrete stirring processes. (cf. [4]). This paper originated a new trend of interest in stochastic flows as dynamical systems. It also motivated the extension of Itô calculus needed for the construction of these flows as solutions of SDE's.

## 2. Instability

We define the Jacobian extension of the flow $\Phi_t$, denoted $\Phi_t^{(1)}$ on $\mathbb{R}^d \times \mathbb{R}^d$, by

$$\Phi_t^{(1)}(x,v) = \left(\Phi_t(x), D_v\Phi_t(x)\right)$$

Denote also $Y_t^x = D\Phi_t(x)$ the Jacobian matrix at $x$, for $\Phi_t$.

For all $x$, we have $Y_\bullet^x \underset{(law)}{\sim} Y_\bullet$ since $W_j^i(t) \underset{(law)}{\sim} D_j W_t^i(x)$. $Y_t$ and $W_j^i(t)$ being the above defined Brownian motions an $SL_d(\mathbb{R})$ and its Lie algebra, with $\alpha = \frac{\mu_2}{d(d+2)}$ ($\mu_2$ is the second moment of $F$.)

**Proposition 1.** ([9] p. 616) $\log(\|Y_t(v)\|)$ *is a diffusion with generator* $\frac{c}{2}\frac{d^2}{dx^2} + \lambda_1 \frac{d}{dx}$, *where* $\lambda_1 = d(d-1)\alpha/2$ *and* $c = (d-1)\alpha$.

The positivity of the Lyapunov exponent $\lambda_1$ implies that $\|Y_t^x(v)\| \to +\infty$ a.s. for any $(x,v)$. The flow is unstable. This intuitively means that "matter" is indefinitely stretched by the flow. But of course, the transformation is not linear. The "matter" is also curved by the flow, and we will now be trying to evaluate this curvature. The situation appears to be very different.

## 3. Curvature

We define the second order extension of the flow denoted $\Phi_t^{(2)}$, on $\left(\mathbb{R}^d\right)^3$, by

$$\Phi_t^{(2)}(x,v,w) = \left(\Phi_t(x), Y_t^x(v), Z_t^x(v,w)\right)$$

with $Z_t^x(v,w) = D_w\phi_t(x) + D_{vv}^2\phi_t(x)$.

*N.B.* If $\Gamma$ is a parametrized curve, and if $\Gamma_t = \Phi_t(\Gamma)$, then

$$\Phi_t^{(2)}\left(\Gamma(0), \Gamma'(0), \Gamma''(0)\right) = \left(\Gamma_t(0), \Gamma_t'(0), \Gamma_t''(0)\right).$$

$Z_t^x$ satisfies the following Itô equation:

$$dZ_t^{x,i}(v,w) = \sum_j Z_t^{x,j}(v,w)\, dW_j^i + dB^i\left(\Phi_t(x), Y_t^x(v), Y_t^x(v)\right)$$

with $B_t^i(x,u,u) = D_{uu}^2 W_t^i(x)$.

Note that this tensor-valued Brownian motion is independent of $D_v W_t^j(x)$ and its law is independent of $x$. Choose $B^i(\cdot,\cdot) \underset{(law)}{\sim} B_t^i(x,\cdot,\cdot)$ and independent of $W_j^i$. Define $Z_t(v,w)$ by:

$$\begin{aligned} Z_0(v,w) &= w \\ dZ_t^i(v,w) &= Z_t^j(v,w)\, dW_j^i + dB_t^i(Y_t(v), Y_t(v)) \\ &= Z_t^j(v,w) \circ dW_j^i + \circ dB_t^i(Y_t(v), Y_t(v)). \end{aligned}$$

Hence,

$$Z_t(v,w) = Y_t\left(w + \int_0^t Y_s^{-1}\ dB_s(Y_s(v), Y_s(v))\right).$$

Set $\Psi_t(v,w) = \big(Y_t(v), Z_t(v,w)\big)$. $\Psi_t$ is the flow of a stochastic differential equation on $\mathbb{R}^d \times \mathbb{R}^d$. For each $t$, $\Psi_t \underset{(law)}{\sim} \Psi_t^{-1}$. If we set $(Y_t^x, Z_t^x) = \Psi_t^x$, $\Psi_t^x \underset{(law)}{\sim} \Psi_t$ for all $x$ and $t$.

We can extend $\Pi$ to $\mathbb{R}^d \times \mathbb{R}^d$ as follows:

$$\Pi(x,y) = \left(\frac{x}{\|x\|}, \frac{y}{\|x\|^2}\right).$$

It is easy to check that $\Pi\Psi_t$ is a flow on $\mathbb{S}^{d-1} \times \mathbb{R}^d$ and that $(\Pi\Psi_t)^{-1} = \Pi\left(\Psi_t^{-1}\right)$. Set $\tau_t(\theta) = \Pi Y_t(\theta) = \frac{Y_t(\theta)}{\|Y_t(\theta)\|}$ and $\rho_t(\theta,w) = \frac{Z_t(\theta,w)}{\|Y_t(\theta)\|^2}$ so that $\Pi\Psi_t = (\tau_t, \rho_t)$.

Define $\rho_t^L = \langle \rho_t, \tau_t\rangle \tau_t$ and $\rho_t^\perp = \rho_t - \rho_t^L$. We can define $\tau_t^x$, $\rho_t^x$, $\rho_t^{x,L}$, $\rho_t^{x,\perp}$ in the same way. For each $x$, they have the same joint distributions as the corresponding variables nonindexed by $x$.

*N.B.* If $x = \Gamma(0), \theta = \Gamma'(0), w = \Gamma''(0)$, $\rho_t^{x,\perp}(\theta,w)$ is the curvature vector of $\Gamma_t$ at $\Phi_t(x) = \Gamma_t(0)$.

**Proposition 2.**

a) $\left\|\rho_t^\perp(\theta,w)\right\|^2$ *is a one-dimensional positive recurrent diffusion with infinitesimal generator:*

$$\begin{aligned} L_x = 2&\left[\alpha(5d-1)x^2 + (d+3)qx\right]\tfrac{\partial^2}{\partial x^2} \\ &+ \left[(d+3)(d-1)q - \alpha\left(d^2 - 9d + 2\right)x\right]\tfrac{\partial}{\partial x} \end{aligned}$$

*with* $\alpha = \mu_2/d(d+2)$ *and* $q = 3\mu_4/d(d+2)(d+4)$.

b) *The asymptotic distribution of* $\left\|\rho_t^{\perp}(\theta,w)\right\|^2$ *is* $cx^v(x+k)^{u-v}$ *with* $k = \frac{q(d+3)}{\alpha(5d-1)}$, $v = \frac{d-3}{2}$, $u = \frac{-(d^2+11d-2)}{10d-2}$, *and c a normalizing constant.*

c) $\left(\left\|\rho_t^{\perp}(\theta,w)\right\|^2, \left\|\rho_t^{L}(\theta,w)\right\|^2\right)$ *is a two dimensional diffusion with generator:*

$$
\begin{aligned}
&L_x + 2\left[2\alpha dxy + (d-1)\alpha y^2 + (d-1)qy\right]\tfrac{\partial^2}{\partial y^2} \\
&\quad + 4\alpha(2d-1)xy\tfrac{\partial^2}{\partial x\partial y} \\
&\quad + \left[2d\alpha x + q(d-1) + \alpha y\left(-d^2+d-2\right)\right]\frac{\partial}{\partial y}.
\end{aligned}
$$

d) $E\left(\left\|\rho_t^{\perp}\right\|^2\right)$ *and* $E\left(\left\|\rho_t^{L}\right\|^2\right)$ *converge towards finite values if* $d \geq 9$ *and otherwise increase to* $+\infty$.

*Proof.* Set

$$
\begin{aligned}
&\theta_t = y_t(\theta), \quad \eta_t = z_t(\theta,\eta) \\
&\xi_t = \|\theta_t\|^2, \quad \lambda_t = \|\eta_t\|^2, \quad \zeta_t = \langle\theta_t,\eta_t\rangle \\
&\varphi_t = \lambda_t/\xi_t^2, \text{ and } \nu_t = \zeta_t^2/\xi_t^3.
\end{aligned}
$$

Then, $\left\|\rho_t^{\perp}\right\|^2 = \varphi_t - \nu_t$ and $\left\|\rho_t^{L}\right\|^2 = \nu_t$. We have

$$
\begin{aligned}
d\theta_t &= dW(\theta_t) \\
d\eta_t &= dW(\eta_t) + dB_t(\theta_t,\theta_t).
\end{aligned}
$$

Set

$$
\begin{aligned}
C(\theta_1,\theta_2) &= \tfrac{1}{t}E\left(W_t(\theta_1)\otimes W_t(\theta_2)\right) \\
\kappa(\theta_1,\theta_2) &= \tfrac{1}{t}E\left(B_t(\theta_1,\theta_1)\otimes B_t(\theta_2,\theta_2)\right).
\end{aligned}
$$

Recall that

$$
\begin{aligned}
&\langle C(\theta_1,\theta_2), \theta_3\otimes\theta_4\rangle \\
&\quad = \alpha\left((d+1)\langle\theta_1,\theta_2\rangle\langle\theta_3,\theta_4\rangle - \langle\theta_1,\theta_3\rangle\langle\theta_2,\theta_4\rangle - \langle\theta_1,\theta_4\rangle\langle\theta_2,\theta_3\rangle\right).
\end{aligned}
$$

Moreover, it is easy to check that if $\langle\theta_1,\theta_2\rangle = 0$,

$$
\begin{aligned}
&\frac{1}{\|\theta_1\|^4\|\theta_2\|^2}\langle\kappa(\theta_1,\theta_1),\theta_2\otimes\theta_2\rangle \\
&\qquad = \mu_4\int\left(u_1^4 - u_2^2u_1^4\right)\sigma(du) = \frac{3(d+3)\mu_4}{d(d+2)(d+4)},
\end{aligned}
$$

and that

$$\frac{1}{\|\theta_1\|^6}\left\langle \kappa\left(\theta_1,\theta_1\right),\theta_1\otimes\theta_1\right\rangle=\mu_4\int\left(u_1^4-u_1^6\right)\sigma\left(du\right)=\frac{3\left(d-1\right)\mu_4}{d\left(d+2\right)\left(d+4\right)}.$$

Hence $\langle\kappa(\theta_1,\theta_1),\theta_2\otimes\theta_2\rangle = q\left[(d+3)\|\theta_1\|^4\|\theta_2\|^2-4\langle\theta_1,\theta_2\rangle^2\theta_1^2\right]$ with $q=3\mu_4/(d(d+2)(d+4))$ and $Tr\left(\kappa(\theta_1,\theta_1)\right)=q(d-1)(d+4)\left\|\theta_1\right\|^4$. Moreover $Tr\left(C(\theta_1,\theta_2)\right)=\alpha(d(d+1)-2)\langle\theta_1,\theta_2\rangle$.

We have the following Itô equations:

$$\begin{aligned}
d\xi_t &= 2\theta_t d\theta_t + Tr\left(C(\theta_t,\theta_t)\right)dt\\
d\lambda_t &= 2\eta_t d\eta_t + Tr\left(\kappa(\theta_t,\theta_t)\right)dt + Tr\left(C(\eta_t,\eta_t)\right)dt\\
d\zeta_t &= \theta_t d\eta_t + \eta_t d\theta_t + Tr\left(C(\theta_t,\eta_t)\right)dt\\
d\langle\xi,\xi\rangle_t &= 4\left\langle C(\theta_t,\theta_t),\theta_t\otimes\theta_t\right\rangle dt\\
d\langle\xi,\lambda\rangle_t &= 4\left\langle C(\theta_t,\eta_t),\eta_t\otimes\theta_t\right\rangle dt\\
d\langle\xi,\zeta\rangle_t &= 2\left\langle C(\theta_t,\eta_t),\theta_t\otimes\theta_t\right\rangle dt + 2\left\langle C(\theta_t,\theta_t),\theta_t\otimes\eta_t\right\rangle dt\\
d\langle\zeta,\zeta\rangle_t &= \left\langle C(\theta_t,\theta_t),\eta_t\otimes\eta_t\right\rangle dt + \left\langle C(\eta_t,\eta_t),\theta_t\otimes\theta_t\right\rangle dt\\
&\quad + 2\left\langle C(\theta_t,\eta_t),\eta_t\otimes\theta_t\right\rangle dt + \left\langle \kappa(\theta_t,\theta_t),\theta_t\otimes\theta_t\right\rangle dt\\
d\langle\lambda,\lambda\rangle_t &= 4\left\langle C(\eta_t,\eta_t),\eta_t\otimes\eta_t\right\rangle dt + 4\left\langle \kappa(\theta_t,\theta_t),\eta_t\otimes\eta_t\right\rangle dt\\
d\langle\zeta,\lambda\rangle_t &= 2\left\langle C(\theta_t,\eta_t),\eta_t\otimes\eta_t\right\rangle dt + 2\left\langle C(\eta_t,\eta_t),\theta_t\otimes\eta_t\right\rangle dt\\
&\quad + 2\left\langle \kappa(\theta_t,\theta_t),\theta_t\otimes\eta_t\right\rangle dt.
\end{aligned}$$

Hence,

$$\begin{aligned}
d\xi_t &= 2\theta_t d\theta_t + \alpha(d^2+d-2)\xi_t dt\\
d\lambda_t &= 2\eta_t d\eta_t + \alpha(d^2+d-2)\lambda_t dt + q(d-1)(d+4)\xi_t^2 dt\\
d\zeta_t &= \theta_t d\eta_t + \eta_t d\theta_t + \alpha(d^2+d-2)\zeta_t dt\\
d\langle\xi,\xi\rangle_t &= 4\alpha(d-1)\xi_t^2 dt,\\
d\langle\xi,\lambda\rangle_t &= 4\alpha(d\zeta_t^2-\xi_t\lambda_t)dt\\
d\langle\xi,\zeta\rangle_t &= 4\alpha(d-1)\xi_t\zeta_t dt\\
d\langle\zeta,\zeta\rangle_t &= 2\alpha\left[(d\xi_t\lambda_t+(d-2)\zeta_t^2)\right]dt + q(d-1)\xi_t^3 dt\\
d\langle\lambda,\lambda\rangle_t &= 4\alpha(d-1)\lambda_t^2 dt + 4q\left[(d+3)\xi_t^2\lambda_t - 4\zeta_t^2\xi_t\right]dt\\
d\langle\lambda,\zeta\rangle_t &= 4\alpha(d-1)\lambda_t\xi_t dt + 2q(d-1)\xi_t^2\zeta_t dt
\end{aligned}$$

$$\begin{aligned}
d\varphi_t &= \frac{d\lambda_t}{\xi_t^2} - \frac{2\lambda_t d\xi_t}{\xi_t^3} - 2\frac{d\langle \lambda, \xi \rangle_t}{\xi_t^3} + \frac{3\lambda_t}{\xi_t^4}\ d\langle \xi, \xi \rangle_t \\
&= -\frac{4\lambda_t}{\xi_t^3}\theta_t d\theta_t + \frac{2\eta_t}{\xi_t^2} d\eta_t + q(d-1)(d+4)dt \\
&\quad - \alpha(d^2 - 11d + 2)\varphi_t dt - 8\alpha d\nu_t dt \\
d\nu_t &= \frac{2\zeta_t}{\xi_t^3} d\zeta_t - 3\zeta_t^2 \frac{d\xi_t}{\xi_t^4} - 6\frac{\zeta_t}{\xi_t^4} d\langle \xi, \zeta \rangle_t + \frac{d\langle \zeta, \zeta \rangle_t}{\xi_t^3} + \frac{6\xi_t^2}{\zeta_t^5} d\langle \xi, \xi \rangle_t.
\end{aligned}$$

Hence,

$$\begin{aligned}
d\nu_t &= \tfrac{\zeta_t}{\xi_t^3}\left(\theta_t d\eta_t + \eta_t d\theta_t\right) - \tfrac{6\zeta_t^2}{\xi_t^4}\theta_t d\theta_t + q(d-1)dt \\
&\quad + 2d\alpha\varphi_t dt - \alpha\left(d^2 - d + 2\right)\nu_t dt
\end{aligned} \tag{1}$$

$$\begin{aligned}
d\langle \varphi, \nu \rangle_t &= \tfrac{2\zeta_t}{\xi_t^5} d\langle \lambda, \zeta \rangle_t - 4\tfrac{\zeta_t\lambda_t}{\xi_t^6} d\langle \zeta, \xi \rangle_t - \tfrac{3\zeta_t^2}{\xi_t^6} d\langle \xi, \lambda \rangle_t + 6\tfrac{\zeta_t^2\lambda_t}{\xi_t^6} d\langle \xi, \xi \rangle_t \\
&= 4\left(q(d-1)\nu_t + \alpha(4d-1)\varphi_t\nu_t - 3\alpha d\nu_t^2\right)dt \\
d\langle \varphi, \varphi \rangle_t &= \tfrac{d\langle \lambda, \lambda \rangle_t}{\xi_t^4} - 4\tfrac{\lambda_t}{\xi_t^5} d\langle \xi, \lambda \rangle_t + 4\tfrac{\lambda_t^4}{\xi_t^6} d\langle \xi, \xi \rangle_t \\
&= 4\left[q\left((d+3)\varphi_t - 4\nu_t\right) + \alpha(5d-1)\varphi_t^2 - 4d\alpha\varphi_t\nu_t\right]dt \\
d\langle \nu, \nu \rangle_t &= 4\tfrac{\zeta_t^2}{\xi_t^6} d\langle \zeta, \zeta \rangle_t - 12\tfrac{\zeta_t^3}{\xi_t^7} d\langle \xi, \zeta \rangle_t + 9\tfrac{\zeta_t^4}{\xi_t^8} d\langle \xi, \xi \rangle_t.
\end{aligned}$$

Hence

$$d\langle \nu, \nu \rangle_t = 4\left[q(d-1)\nu_t + (d-1)\alpha\nu_t^2 + 2\alpha d\varphi_t\nu_t\right]dt \tag{2}$$

$$d\langle \varphi - \nu, \varphi - \nu \rangle_t = 4\left[q(d+3)(\varphi_t - \nu_t) + (5d-1)(\varphi_t - \nu_t)^2\right]dt. \tag{3}$$

Moreover,

$$\begin{aligned}
&d\varphi_t - d\nu_t \\
&= d(\text{local martingale}) - \alpha\left(d^2 - 9d + 2\right)(\varphi_t - \nu_t)\,dt + q(d-1)(d+3)dt
\end{aligned} \tag{4}$$

$$\begin{aligned}
&d\varphi_t - 4d\nu_t \\
&\qquad = d(\text{local martingale}) + qd(d-1) - \alpha\left(d^2 - 3d + 2\right)(\varphi_t - 4\nu_t)\,dt
\end{aligned} \tag{5}$$

Assertions a) and c) follow from formulas (1)–(4). a) implies b) by standard calculations on one dimensional diffusions. Obtaining d) is easy by taking expectations in (4) and (5), since we get a linear system of differential equations that can be explicitly solved. We need to check that for finite $t$, $E(\varphi_t^n)$ and $E(\nu_t^n)$, are bounded for all integer $n$. This will imply that the local martingales in (4) and (5) are square integrable up to any finite time.

But $E\left(\xi_t^{-n}\right)$ can be evaluated since the law of $\xi_t$ is known (cf. Prop. 1) and from the SDE defining $\eta_t$, it is clear that $E\left(\lambda_t^n\right) < \infty$.

## 4. Unstable directions

Define $\hat{\Phi}_t = \Phi_t^{-1}$ and $\hat{Y}_t^x = \left[Y_t^x\right]^{-1}$ the inverse Jacobian matrix of $\Phi_t$ at $x$ or the Jacobian of $\hat{\Phi}_t$ at $\Phi_t(x)$. In each point $x$, we define a random unstable direction as follows:

**Proposition 3.** *$P(dw) \otimes \sigma(d\theta)$ a.s., the direction of $\hat{Y}_t^x(\theta)$ converges towards a limiting direction independent of $\theta$. (Recall $\sigma$ is the normalized lebesque measure on $\mathbb{S}^{d-1}$.)*

*Proof.* We can replace $Y^x$ by $Y$. $\Pi Y_t$ is a Brownian flow on $\mathbb{S}^{d-1}$. In such a situation, it is well known, and easy to check that $\left(\Pi Y_t\right)^{-1} \sigma(d\theta) \underset{(def)}{=} \sigma_t(d\theta)$ is a measure valued martingale (cf [11] for example.)

Here we have explicitly $\frac{d\sigma_t(\theta)}{d\sigma(\theta)} = \frac{1}{\|Y_t(\theta)\|^d}$, which is a diffusion with generator $\alpha\, d^2\,(d-1)\, x^2\, \frac{d^2}{dx^2}$, hence it is a martingale.

The projection on the direction of $\hat{Y}_t(\theta)$ is

$$\frac{\hat{Y}_t(\theta)\,\hat{Y}_t(\theta)^*}{\left\|\hat{Y}_t(\theta)\right\|^2} = \widehat{\Pi Y_t}(\theta)\,\widehat{\Pi Y_t}(\theta)^* \underset{(def)}{=} A_t^\theta.$$

The average of these projectors under $\sigma(d\theta)$ is a random matrix

$$\begin{aligned} A_t &= \int \widehat{\Pi Y_t}(\theta) \quad \widehat{\Pi Y_t}(\theta)^*\,\sigma(d\theta) \\ &= \int \eta\eta^*\sigma_t(d\eta)\,. \end{aligned}$$

Hence $A_t$ is a martingale which is clearly bounded. Let $A_\infty$ be its limit a.s. To check that $A_\infty$ is a projector and that $A_\infty$ is $\lim_{t\to\infty} A_t^\theta \quad \sigma(d\theta)$ a.s., it is enough to check that $\iint Tr\Big(\big(A_t^\theta - A_t^{\theta'}\big)^2\Big)\sigma(d\theta)\,\sigma(d\theta') \to 0$ a.s. as $t \uparrow \infty$.

$$Tr\Big(\big(A_t^\theta - A_t^{\theta'}\big)^2\Big) = 1 - \frac{\langle Y_t(\theta), Y_t(\theta')\rangle^2}{\|Y_t(\theta)\|^2\,\|Y_t(\theta')\|^2} = \frac{\|Y_t(\theta)\wedge Y_t(\theta')\|^2}{\|Y_t(\theta)\|^2\,\|Y_t(\theta')\|^2}.$$

But $\log\|Y_t(\theta)\wedge Y_t(\theta')\|/t \;\sim\; \lambda_1+\lambda_2$ if $\theta \neq \theta'$, where $\lambda_2 = d(d-3)\,\alpha/2$ (cf. [9], p. 616).

Hence

$$\log\left[Tr\left(\left(A_t^\theta - A_t^{\theta'}\right)^2\right)\right]/t \sim\; 2(\lambda_2-\lambda_1)$$
$$\sim -\,2d\alpha \quad \text{a.s.}$$

*N.B.* This result is in fact a special case of general results, well known for products of random matrices.

## 5. Longitudinal energy and Laplacian

A priori, the field of random directions defined by the projectors $A_\infty^x$ does not have a smooth version, so we cannot integrate it. However, the study of deterministic dynamical systems (Sinai, Anosov, Pesin, Ruelle, etc. $\cdots$) suggests integral curves do exist, are smooth and that the foliation they define is absolutely continuous, i.e. maps lebesque measure on one transversal onto an absolutely continuous measure on another transversal. Such results have been obtained by Ledrappier and Young [6] for a class of stochastic flows on compact manifolds.

Their techniques require an important background in dynamical systems.

Here, we are not directly trying to study such a foliation. But the result we are going to obtain are very closely related to such properties as smoothness and absolute continuity.

Intuitively, the longitudinal diffusion we are going to construct is a diffusion on these integral curves. Its tranversal drift, whose law shall be determined explicitly, can be interpreted as the curvature of the foliation.

Note that $\left[\Phi_t^{(i)}\right]^{-1} = \left[\Phi_t^{-1}\right]^{(i)}$ for $i = 1, 2$. Set $(\Psi_t^x)^{-1} = \hat{\Psi}_t^x$ and define $\hat{Z}_t^x$ such that $\hat{\Psi}_t^x(u,v) = \left(\hat{Y}_t^x(u), \hat{Z}_t^x(u,v)\right)$. For fixed $t$, $\left(\hat{Y}_t^x, \hat{Z}_t^x\right) \underset{(law)}{\sim} (Y_t, Z_t)$. If $\Gamma(s)$ is a curve such that $\Gamma(0) = 0$, $\Gamma'(0) = u$, $\Gamma''(0) = v$, and $\hat{\Gamma}_t = \phi_t^{-1}(\phi_t(x) + \Gamma)$, then $\hat{Y}_t^x(u) = \Gamma_t'(0)$ and $\hat{Z}_t^x(u,v) = \hat{\Gamma}_t''(0)$. Set, as before,

$$\Pi\left(\hat{Y}_t^x, \hat{Z}_t^x\right) = (\hat{\tau}_t^x, \hat{\rho}_t^x), \quad \hat{\rho}_t^{x,L} = \langle \hat{\rho}_t^x, \hat{\tau}_t^x \rangle \hat{\tau}_t^x, \quad \hat{\rho}_t^{x,\perp} = \hat{\rho}_t^x - \hat{\rho}_t^{x,L}.$$

$\hat{\rho}_t^{x,\perp}$ is the curvature vector of $\hat{\Gamma}_t$ at $x$.

Assume now that $\Gamma$ is a straight line, with $\Gamma'(s) = \theta$ for all $s$. Then $\hat{Y}_t^{\hat{\Gamma}_t(s)}(\theta) = \frac{d}{ds}\hat{\Gamma}_t(s)$. Hence, $\hat{\rho}_t^{x,\perp} = \frac{1}{\|\hat{\Gamma}_t'(0)\|} \cdot \frac{d}{ds}\hat{\tau}^{\hat{\Gamma}_t(s)}|_{s=0} = \sum\limits_i \hat{\tau}_t^{x,i} \frac{d}{dx_i}\hat{\tau}_t^x = \nabla_{\hat{\tau}_t}\hat{\tau}_t(x)$,

$$\begin{aligned}\left\langle \hat{\tau}_t, \hat{\rho}_t^{x,L} \right\rangle = \frac{\left\langle \hat{\Gamma}_t''(0), \hat{\tau}_t^x \right\rangle}{\|\Gamma_t'(0)\|^2} &= -\frac{d}{ds}\left(\frac{1}{\|\Gamma_t'(s)\|}\right)_{s=0} \\ &= -\left[\sum_i \hat{Y}_t^{x,i} \frac{d}{dx_i}\left(\frac{1}{\left\|\hat{Y}_t^x\right\|}\right)\right] = -D_{\hat{Y}_t}\left(\frac{1}{\left\|\hat{Y}_t\right\|}\right).\end{aligned}$$

Since $\phi_t^{-1}$ is measure preserving, $\operatorname{div}\left(\hat{Y}_t^x(\theta)\right) = 0$ for all $\theta$. Hence $D_{\hat{Y}_t}\left(\frac{1}{\|\hat{Y}_t\|}\right) = \operatorname{div}(\hat{\tau}_t)$ so that $\hat{\rho}_t^{x,L} = -\operatorname{div}(\hat{\tau}_t^x)\,\hat{\tau}_t^x$. Finally, we get the following integration by parts formula.

For any $f, g \in C^2(\mathbb{R}^d)$, with compact support,

$$\begin{aligned}\int D_{\hat{\tau}_t} f(x) D_{\hat{\tau}_t} g(x)dx = &-\int g(x) D^2_{\hat{\tau}_t\hat{\tau}_t} f(x)dx \\ &-\int g(x)\left[D_{\hat{\rho}_t^\perp} f(x) - D_{\hat{\rho}_t^L} f(x)\right]dx.\end{aligned}$$

Integrating both sides by $\sigma(d\theta)$ we get

(6)

$$\begin{aligned}\int A_t^{ij}(x) D_i f(x) D_j g(x)dx = &-\int A_t^{ij} D^2_{ij} f(x) g(x)dx \\ &-\int g(x)\left[D_{b_t^\perp} f(x) - D_{b_t^L} f(x)\right]dx,\end{aligned}$$

where

$$A_t^{ij}(x) = \int \hat{\tau}_t^{x,i}\hat{\tau}_t^{x,j}(\theta)\sigma(d\theta), \quad b_t^{\perp(L)}(x) = \int \rho_t^{x,\perp(L)}(\theta,0)\sigma(d\theta).$$

Since $Y^x \underset{(law)}{\sim} Y$, we have seen that $A_t^{ij}(x)$ is a bounded martingale. Let $A_\infty(x)$ be its a.s. limit. The first member of (6) converges towards

$$\mathcal{E}_L(f,g) \underset{(def)}{=} \int A_\infty^{ij}(x)D_i f(x)D_j g(x)dx.$$

We call $\mathcal{E}_L(\rho, f)$ the longitudinal energy of $f$.

**Theorem.** *Assume $d \geq 9$. (a) For all $x, b_t^L(x)$ and $b_t^\perp(x)$ are $L^2$-converging martingales with $A_\infty b_\infty^L = b_\infty^L$ and $A_\infty b_\infty^\perp = 0$. (b) The law of $\|b_\infty^\perp(x)\|^2$ has the density $cx^v(x+b)^{u-v}$. (c) $\mathcal{E}_L$ is a.s. closable and the associated generator is the Friedrich extension of*

$$\Delta_\omega = \sum_{ij} A_\infty^{ij} D_{ij}^2 + D_{b_\infty^\perp} - D_{b_\infty^L}.$$

*It is a local Dirichlet form. It generates a self adjoint-Markovian semigroup on $L^2(dx)$ and a continuous Markov process (which we can call the longitudinal diffusion).*

*Remark.* The operator $\tilde{\Delta}_\omega = \Delta_\omega + 2D_{b_\infty^L}$ is not symmetric with respect to lebesgue measure but has a nice quasi-invariance property:

$$\tilde{\Delta}_\omega (f \circ \phi_t)(x) = Tr\left(Y_t A_\infty Y_t^*\right) \tilde{\Delta}_{\theta_t\omega} f\left(\Phi_t(x)\right)$$

$\theta_t$ being the canonical shift: $\Phi_t(\theta_s\omega) = \Phi_{t+s} \circ \Phi_s^{-1}(\omega)$.

*Proof.* a) Note first that $\hat{Z}_t^x(\theta,0) = -\hat{Y}_t^x Z_t^x\left(\hat{Y}_t^x(\theta),0\right)$. Hence,

$$\hat{\rho}_t^x(\theta,0) = -\hat{Y}_t^x Z_t^x\left(\hat{\tau}_t^x(\theta),0\right)$$

$$b_t^L = \int \frac{\left\langle \eta, \hat{Y}_t^x Z_t^x(\eta,0)\right\rangle}{\|Y_t^x(\eta)\|^d} \eta\sigma(d\eta)$$

$$b_t \underset{(def)}{=} b_t^\perp + b_t^L = \int \frac{\hat{Y}_t^x Z_t^x(\eta,0)}{\|Y_t^x(\eta)\|^d}\sigma(d\eta).$$

Set

$$M_t^{x,\eta} = \hat{Y}_t^x Z_t^x(\eta,0) = D\Phi_t^{-1} D^2_{\eta\eta}\Phi_t(x)$$
$$N_t^{x,\eta} = \|Y_t^x(\eta)\|^{-d} = \|D_\eta \Phi_t(x)\|^{-d}.$$

$N_t^{x,\eta}$ is a martingale as well as $M_t^{x,\eta}$, since

$$dM_t^{x,\eta} = \hat{Y}_t^x dB_t(\Phi_t(x), Y_t^x(\eta), Y_t^x(\eta)).$$

Moreover, one checks $\langle M^{x,\eta}, N^{x,\eta}\rangle_t = 0$, so that $b_t^L$ and $b_t^N$ are martingales. Besides, since

$$\mathrm{E}\left(\|\hat{\rho}_t^x(\theta,0)\|^2\right) = \mathrm{E}\left(\|\rho_t(\theta,0)\|^2\right)$$

these martingales are $L^2$ bounded for $d \geq 9$.

To prove that $A_\infty b_\infty^\perp = 0$, note that

$$\begin{aligned}
E\left(\|A_\infty b_\infty^\perp\|\right)^2 &= \lim_{t\to\infty} E\left(\|A_t b_t^\perp\|\right)^2 \\
&\leq \liminf_{t\to\infty} \int d\theta\, d\theta' E\left(\left\|A_t^{\theta'}\ \hat{\rho}_t^\perp(\theta)\right\|\right)^2 \\
&= \liminf_{t\to\infty} \int d\theta d\theta' E\left(\left\|\left(A_t^{\theta'} - A_t^\theta\right)\rho_t^\perp(\theta)\right\|\right)^2 \\
&\leq \liminf_{t\to\infty} \int d\theta d\theta' E\left(\left\|A_t^{\theta'} - A_t^\theta\right\|^2\right) E\left(\|\rho_t^\perp(\theta)\|^2\right) \\
&= 0.
\end{aligned}$$

Similar estimates yields $A_\infty b_\infty^L = b_\infty^L$. c) follows directly from a) and the above integration by parts formula (cf. [3]). b) We can work with $\Psi_t$ and $\hat{\Psi}_t = \Psi_t^{-1}$ instead of $\Psi_t^x$. $\Psi_t$ induces naturally a flow $\tilde{\Psi}_t$ on $\mathbb{S}^{d-1}/\pm 1 \times \mathbb{R}^d$

$$\tilde{\Psi}_t(\pm\theta, y) = \left(\pm\frac{Y_t(\theta)}{\|Y_t(\theta)\|}, \rho_t^\perp(\pm\theta, y)\right).$$

The induced one-point diffusion on $\mathbb{S}^{d-1}/\pm 1 \times \mathbb{R}^d$ has $\tilde{\sigma}(d(\pm\theta))\,\sigma_{\pm\theta}(\eta)h(r)dr = m$ as invariant probability with

$$\begin{aligned}
\tilde{\sigma}(d(\pm\theta)) &= \text{ lebesgue measure on } \mathbb{S}^{d-1}/\pm 1 \\
\sigma_{\pm\theta}(d\eta) &= \text{ lebesgue measure on } \{\eta \in S_{d-1}\ \langle \pm\theta, \eta\rangle = 0\} \\
h(r) &= 2cr^{2v+1}(r^2+k)^{2(u-v)} \\
r &= \|y\|, \quad \eta = r^{-1}y.
\end{aligned}$$

Let $\mu_\omega\left(d(\pm\theta), dy\right)$ be the limit of the measure valued martingale $\left[\tilde{\Psi}_t\right]^{-1}(m)$. We can check that

$$\begin{aligned} b_\infty^\perp &= \lim_{t\to\infty} \int d\theta \hat{\rho}_t(\theta, 0) \\ &= \int y\mu_\omega(d\theta, dy). \end{aligned}$$

Since $\|\tau_t(\theta) \wedge \tau_t(\theta')\| \to 0$ a.s. and

$$\|\Psi_t(\theta, y) - \Psi_t(\theta, y')\| \leq \frac{\|Y_t(y - y')\|}{\|Y_t(\theta)\|^2} \to 0 \ a.s.$$

$\mu_\omega\left(d(\pm\theta), dy\right)$ is a Dirac measure $\epsilon_{(\pm\theta_\omega, y_\omega)}$. Since $\mathrm{E}(\mu_\omega) = m$, $(\pm\theta_\omega, y_\omega)$ is distributed according to $m$. Since $b_\infty^\perp = y_\omega$, we are done.

*Remarks.*

a) The condition $d \geq 9$ should be removed by using sharper techniques for convergence.
b) The same study could be done for non measure - preserving isotropic flows. The intergration by parts formula should be given with respect to $\mu_+^\omega$, the sample measure of the statistical equilibrium (cf. [9], p. 619).
c) For $d \geq 2n$, similar results should be obtained for $n$-dimensional submanifolds.
d) There results have been announced in a note [12].

## References

[1] Baxendale, P., *Brownian motion on the diffeomorphism group I*, Compositio Mathematica **53** (1984), pp. 19–50.

[2] Baxendale, P. and Harris, T.E., *Isotropic stochastic flows*, Annals of Probab. **14** (1986), pp. 1155–1179.

[3] Fukushima, M., "Dirichlet Forms and Markov Processes," North Holland-Kodansha, 1980.

[4] Harris, T.E., *Brownian motions on the homeomorphisms of the plane*, Ann. Probab. **9** (1981), pp. 232–254.

[5] Kunita, H., "Stochastic Flows and S.D.E.'s," Cambridge University Press, 1991.

[6] Itô, K., *Isotropic random currents*, "Proc. Third Berkeley Symp. 2," Univ. Calif. Press, Berkeley, 1956.

[7] Ledrappier, F. and Young, L-S., *Entropy formula for random transformations*, Prob. Th. Rel. Fields **80** (1988), pp. 217–240.

[8] Le Jan, Y., *Flots de diffusion dans* $\mathbb{R}^d$, C. R. Acad Sci. Paris Serie I **294** (1982), pp. 697–699.

[9] Le Jan, Y., *On isotropic Brownian motions*, Zeit. für Wahrsch **70** (1985), pp. 609–620.

[10] Le Jan, Y. and Watanabe, S., *Stochastic flows of diffeomorphisms*, "Proc. of the Taniguchi Symposium, (edited by K. Itô)," North Holland, 1984.

[11] Le Jan, Y., *Equilibre statistique pour les produits de difféomorphismes aléatoires indépendants*, Ann. Inst. Henri Poincaré **23** (1987), pp. 111–120.

[12] Le Jan, Y., *Propriétés asymptotiques des flots browniens isotropes*, C. R. Acad Sci. Paris Serie I **309** (1989), pp. 63–65.

[13] Yaglom, A., *Some classes of randonm fields in n-dimensional space*, Theory Probab. Appl. **2** (1957), pp. 273–320.

Laboratoire de Probabilités (CNRS)
75230 Paris Cedex 05
France

# Ergodic Properties of Nonlinear Filtering Processes

Hiroshi Kunita

## Introduction

Let $x_t$ be a temporally homogeneous Markov process with state space $S$, called a *system process* in this paper. Suppose that we want to observe the sample path $x_t$, but what we can observe is a stochastic process $Y_t$ of the form

$$Y_t = \int_0^t h(x_s)dt + N_t, \tag{0.1}$$

where $h$ is a continuous function on $S$ and $N_t$ is a standard Brownian motion independent of $x_s$. The filtering of the system process based on the observation data $Y_t$ is defined by a conditional distribution

$$\pi_t(A) = P(x_t \in A|\mathcal{G}_t), \tag{0.2}$$

where $A$ is a Borel subset of $S$ and

$$\mathcal{G}_t = \bigcap_{\varepsilon>0} \sigma\left(Y_s; s \leq t+\varepsilon\right). \tag{0.3}$$

In the paper [5], the author proved that the filtering process $\pi_t$ is a temporally homogeneous Markov process with state space $\mathcal{M}_1$ provided that $x_t$ is a Feller process with compact state space $S$. Here $\mathcal{M}_1$ is the set of all probability measures on $S$. In this paper we study the case that the state space $S$ is noncompact. We will, however, restrict our attention to the case that the system process is governed by a stochastic differential equation on a Euclidean space. We will show in Section 1 that not only the filtering process $\pi_t$ but also the pair process $(x_t, \pi_t)$ is a temporally homogeneous Markov process. This fact permits us to study the ergodic property of the pair process. In Section 2, we will discuss the weak convergence of the transition probabilities of the pair process in the case that the system process has an invariant probability. Applying this, one can obtain the limit of the moments of the filtering error:

$$\lim_{t\to\infty} E\left[\left|f(x_t) - \int \pi_t(dx) f(x)\right|^p\right]. \tag{0.4}$$

It may be interesting to compare these results with the linear filter. If the system process is a Gaussian Markov process and the function $h(x)$ appearing in (0.1) is a linear function $Hx$, then the filter $\pi_t$ is a Markov process with values in Gaussian distributions. Its mean $\hat{x}_t = E[x_t|\mathcal{G}_t]$ is governed by Kalman's linear stochastic differential equation. Its covariance $P_t$ is deterministic and is governed by a suitable matrix Riccati equation. See Section 2 for details. Hence the ergodic property of the pair process $(x_t, \pi_t)$ can be obtained from the ergodic property of the triple process $(x_t, \hat{x}_t, P_t)$. Especially the asymptotic behavior of the error covariance $P_t$ of the filtering can be analyzed through the matrix Riccati equation governing $P_t$. Unfortunately there is no counterpart of this theory in nonlinear filtering theory. Namely, there is no finite dimensional system of differential equations governing the filtering error covariance. In Section 3, we obtain two Riccati differential equations: The solution of one equation majorizes the error covariance $E[|x_t - \hat{x}_t|^2]$ and the solution of the other equation minorizes $E[|x_t - \hat{x}_t|^4]^{1/2}$. Then the limits of these two solutions as $t \uparrow \infty$ give the upper and lower estimates of the limits of the variance and the fourth moment of the error, respectively.

## 1. Filtering process and the associated Markov process

Let $\sigma_j(x) = \left(\sigma_j^1(x), \dots, \sigma_j^d(x)\right), j = 1, \dots, n$ be $C_b^2$-functions on $\mathbf{R}^d$ and $\sigma_0(x) = \left(\sigma_0^1(x), \dots, \sigma_0^d(x)\right)$ be a $C^1$-function on $\mathbf{R}^d$ with bounded derivatives. Let $B_t = \left(B_t^1, \dots, B_t^n\right)$ be an $n$-dimensional standard Brownian motion such that $B_0 = 0$ defined on the probability space $(\Omega, \mathcal{F}, P)$. Consider Stratonovich's stochastic differential equation in $\mathbf{R}^d$:

$$x_t = x_0 + \sum_{j=1}^{n} \int_0^t \sigma_j(x_s) \circ dB_s^j + \int_0^t \sigma_0(x_s)\, ds, \tag{1.1}$$

where $x_0$ is a random variable independent of $B_t$. It has a unique solution $x_t^{(x_0)}$. The superior $(x_0)$ is often omitted and is denoted by $x_t$. It is called a *system process.*

The system process is a temporally homogeneous Markov process. Let $x_t^{(x)}$ be the solution with the initial condition $x_0 = x$, where $x$ is a fixed point of $\mathbf{R}^d$. Set $P_t(x, A) = P(x_t^{(x)} \in A)$. It is the transition probability of the system process. Further, $P_t$ defined by $P_t f(x) = \int P_t(x, dy) f(y)$ maps $C_\infty(\mathbf{R}^d)$ into itself, where $C_\infty(\mathbf{R}^d)$ is the Banach space of continuous functions on $\mathbf{R}^d$ vanishing at infinity. Define vector fields $X_j, j = 0, \ldots, n$ by $X_j = \sum_i \sigma_j^i(x)\left(\partial/\partial x^i\right)$ and a second order linear partial differential operator $L$ by

$$L = \tfrac{1}{2} \sum_{j=1}^{n} X_j^2 + X_0. \tag{1.2}$$

Then $L$ is the infinitesimal generator of the system process, i.e.,

$$P_t f(x) = f(x) + \int_0^t P_s L f(x) ds \tag{1.3}$$

is satisfied for any $C^2$-function $f$ with bounded derivatives.

Now let $h(x) = (h^1(x), \ldots, h^m(x)), x \in \mathbf{R}^d$ be a $C^3$-function with bounded derivatives. Let $\mathcal{M}_1$ be the set of all probabilities on $\mathbf{R}^d$. In the sequel, for $\nu \in \mathcal{M}_1$ and real measurable function $f$, we use the notation

$$\nu(f) = \int f(x) \nu(dx)$$

if the integral is well defined. We shall consider a stochastic differential equation of an $\mathcal{M}_1$-valued process

$$\begin{aligned} \nu_t(f) = \nu_0(f) &+ \int_0^t \nu_s(Lf) ds \\ &+ \sum_i \int_0^t \left(\nu_s(fh^i) - \nu_s(f)\nu_s(h^i)\right) \\ &\qquad \times \left(h^i(x_s) ds - \nu_s(h^i) ds + dN_s^i\right), \end{aligned} \tag{1.4}$$

where $N_t = \left(N_t^1, \dots, N_t^m\right)$ is a standard $m$-dimensional Brownian motion independent of $B_t$, and $\nu_0$ is an $\mathcal{M}_1$-valued random variable independent of $N_t$ and $B_t$. The pair of equations (1.1) and (1.4) can be regarded as a closed system of equations.

Equation (1.4) is originated from the nonlinear filtering problem. Suppose that the observation process $Y_t$ is defined by (0.1). Then the filtering process $\pi_t$ defined by (0.2) satisfies the following stochastic partial differential equation called Kushner-Stratonovich equation or $FKK$ equation.

$$
\begin{aligned}
\pi_t(f) &= \pi_0(f) + \int_0^t \pi_s(Lf)ds \\
&\quad + \sum_i \int_0^t \left(\pi_s(fh^i) - \pi_s(f)\pi_s(h^i)\right) \\
&\qquad \times \left(dY_s^i - \pi_s(h^i)ds\right).
\end{aligned} \tag{1.5}
$$

See Fujisaki-Kallianpur-Kunita [1]. Therefore if the distribution of the initial random variable $x_0$ coincides with $\nu_0$ in equation (1.4), then the filtering $\pi_t$ is a solution of equation (1.4) since $h^i(x_s)ds + dN_s^i$ coincides with the observation $dY_s^i$. More generally if $\nu_0$ is a random variable independent of $B_t$ and $N_t$ and satisfies

$$
E\left[f\left(x_0\right)|\sigma\left(\nu_0\right)\right] = \nu_0\left(f\right) \qquad \forall f \in C_\infty\left(\mathbb{R}^d\right), \tag{1.6}
$$

the modified filtering $\hat{\pi}_t(f) \equiv E\left[f\left(x_t\right)|\mathcal{G}_t \vee \sigma\left(\nu_o\right)\right]$ is also a solution of equation (1.4). However if $x_0$ and $\nu_0$ are not related by (1.6), equation (1.4) is not an equation for the filtering.

Now we will first discuss the existence and the uniqueness of the solution of equation (1.4).

**Theorem 1.1.** *Let $(x_0, \nu_0)$ be an $\mathbb{R}^d \times \mathcal{M}_1$ -valued random variable independent of $(B_t, N_t)$. Then the pair of equations (1.1) and (1.4) has a unique solution with values in $\mathbb{R}^d \times \mathcal{M}_1$.*

*Proof.* The theorem is shown in Kunita [7] essentially. We quote here its outline for the convenience of the reader. Instead of equation (1.4), we consider a linear stochastic partial differential equation for a measure-valued process $\rho_t$ as an intermediate equation

$$\rho_t(f) = \rho_0(f) + \int_0^t \rho_s(Lf)ds + \sum_i \int_0^t \rho_s(h^i f)dY_s^i. \tag{1.7}$$

Here $\rho_t$ is a stochastic process with values in $\mathcal{M}$, where $\mathcal{M}$ is the set of all Radon measures on $\mathbb{R}^d$, $Y_t = \int_0^t h\left(x_s^{(x_0)}\right) ds + N_t$ and $x_t^{(x_0)}$ is the solution of equation (1.1). Using the Stratonovich integral, the above is equivalent to

$$\rho_t(f) = \rho_0(f) + \int_0^t \rho_s(Af)ds + \sum_i \int_0^t \rho_s(h^i f) \circ dY_s^i, \tag{1.8}$$

where $A = L - (1/2)\sum_i (h^i)^2$. We will consider the last term involving the stochastic integrals as a perturbation adjoined to the partial differential equation. Let $\eta_t(\cdot)(x)$ be an $\mathcal{M}$-valued process satisfying the equation

$$\eta_t(f)(x) = f(x) + \sum_i \int_0^t \eta_s\left(h^i f\right) \circ dY_s^i.$$

The above $\eta_t$ is unique and is represented by

$$\eta_t(f)(x) = f(x) \exp\left\{\sum_i h^i(x) Y_t^i\right\}.$$

Set $\eta_t^{-1}(f)(x) = f(x) \exp\left\{-\Sigma_i h^i(x) Y_t^i\right\}$ and define $A_\omega(t) f = \eta_t A \eta_t^{-1} f$. For almost all $\omega$, it is a second order elliptic linear partial differential operator. Consider the partial differential equation for $\zeta_t$ with values in $\mathcal{M}$.

$$\frac{\partial \zeta_t(f)}{\partial t} = \zeta_t\left(A_\omega(t) f\right), \qquad \zeta_0(f) = \rho_0(f), \qquad \forall f \in C_0^\infty\left(\mathbb{R}^d\right).$$

For almost all $\omega$, it has a unique solution $\zeta_t(f)$. Then $\rho_t(f) \equiv \zeta_t\left(\eta_t(f)\right)$ is the unique solution of equation (1.8).

Now let $\rho_t$ be a solution of equation (1.8). Then we can show using Itô's formula that $\nu_t(f) \equiv \rho_t(f)/\rho_t(1)$ is a solution of equation (1.4). Conversely let $\nu_t$ be a solution of equation (1.4). We define

$$\rho_t(f) = \nu_t(f) \exp\left\{\sum_i \int_0^t \nu_s(h^i)\, dY_s^i - \tfrac{1}{2} \sum_i \int_0^t \nu_s(h^i)^2\, ds\right\}.$$

Then again using Itô's formula, we can show that $\rho_t$ is a solution of equation (1.8). Therefore the existence and uniqueness of the solution of equation (1.4) can be reduced to those of equation (1.8).

**Theorem 1.2.** *The pair of solutions $\left(x_t^{(x_0)}, \nu_t^{(x_0,\nu_0)}\right)$ of equations (1.1) and (1.4) has the following property:*

(i) *It is a Markov process with transition probabilities*

$$\Pi_t(x, \nu, A \times \Gamma) \equiv P\left(x_t^{(x)} \in A, \nu_t^{(x,\nu)} \in \Gamma\right), \tag{1.9}$$

*where $(x, \nu)$ is a fixed element of $\mathbf{R}^d \times \mathcal{M}_1$.*

(ii) *If the initial data $x_0$ and $\nu_0$ are related by (1.6), then $\nu_t^{(x_0,\nu_0)}$ is a Markov process with transition probabilities*

$$\tilde{\Pi}_t(\nu, \Gamma) \equiv \int \Pi_t\left(x, \nu, \mathbf{R}^d \times \Gamma\right) \nu(dx). \tag{1.10}$$

*Further it coincides with the modified filtering $E\left[f(x_t)|\mathcal{G}_t \vee \sigma(\nu_0)\right]$.*

*Proof.* Let $s > 0$ and $\theta_s$ be the shift operator on $\Omega$ such that $B_t(\theta_s\omega) = B_{t+s}(\omega)$ and $N_t(\theta_s\omega) = N_{t+s}(\omega)$ holds for any $t$ a.s. Then we have the relations $x_{t+s}^{(x_0)}(\omega) = x_t^{(x_s)}(\theta_s\omega)$ and $\nu_{t+s}^{(x_0,\nu_0)}(\omega) = \nu_t^{(x_s,\nu_s)}(\theta_s\omega)$ by the uniqueness of the solution of equations (1.1) and (1.4) where $x_s = x_s^{(x_0)}$ and $\nu_s = \nu_s^{(x_0,\nu_0)}$. Set

$$\mathcal{F}_s = \sigma(B_u, N_u : u \le s) \vee \sigma(x_0, \nu_0).$$

Since $x_t^{(\cdot)}(\theta_s\omega)$ and $\nu_t^{(\cdot)}(\theta_s\omega)$ are independent of $\mathcal{F}_s$, we have

$$\begin{aligned}P\left(x_{t+s}^{(x_0)} \in A, \nu_{t+s}^{(x_0,\nu_0)} \in \Gamma|\mathcal{F}_s\right) &= P\left(x_t^{(x)} \in A, \nu_t^{(x,\nu)} \in \Gamma\right)|_{x=x_s,\nu=\nu_s} \\ &= \Pi_t\left(x_s^{(x_0)}, \nu_s^{(x_0,\nu_0)}, A \times \Gamma\right).\end{aligned}$$

Therefore the pair is a Markov process with transition probabilities $\Pi_t$.

Suppose that the initial data $x_0$ and $\nu_0$ are related by (1.6). Then the modified filter $\hat{\pi}_t(f)$ satisfies equation (1.5). By the pathwise uniqueness of the solution of equation (1.4), we have $\nu_t^{(x_0,\nu_0)} = \hat{\pi}_t$. It satisfies

$$P\left(\nu_{t+s}^{(x_0,\nu_0)} \in \Gamma|\mathcal{F}_s\right) = \Pi_t\left(x_s^{(x_0)}, \nu_s^{(x_0,\nu_0)}, \mathbb{R}^d \times \Gamma\right).$$

Take the conditional expectation of both sides with respect to $\mathcal{G}_t \vee \sigma(\nu_0)$. Then we have

$$P\left(\nu_{t+s}^{(x_0,\nu_0)} \in \Gamma|\mathcal{G}_s \vee \sigma(\nu_0)\right) = \tilde{\Pi}_t\left(\nu_s^{(x_0,\nu_0)}, \Gamma\right).$$

This proves that $\nu_t^{(x_0,\nu_0)}$ is a Markov process with transition probabilities $\tilde{\Pi}_t$. The proof is complete.

**Theorem 1.3.** *Both of $\Pi_t$ and $\tilde{\Pi}_t$ map the space of bounded continuous functions into itself.*

*Proof.* Let $(x_n, \nu_n) \in \mathbb{R}^d \times \mathcal{M}_1$ be a sequence converging to $(x, \nu)$. It is enough to show that the solution $\left(x_t^{(x_n)}, \nu_t^{(x_n,\nu_n)}(f)\right)$ converges to $\left(x_t^{(x)}, \nu_t^{(x,\nu)}(f)\right)$ a.s. Since the convergence $x_t^{(x_n)} \to x_t^{(x)}$ is well known, we shall consider the convergence of the second components only.

For this purpose, we represent the solution of equation (1.4) using another probability space $(W, \mathcal{B}, Q)$. Let $W_t$ be a standard $n$-dimensional Brownian motion on $(W, \mathcal{B}, Q)$. Consider a stochastic differential equation:

$$\varphi_t = y + \sum_j \int_0^t \sigma_j(\varphi_s) \circ dW_s^j + \int_0^t \sigma_0(\varphi_s)\, ds.$$

Denote the solution by $\varphi_t(y)$. Then the solution $\nu_t^{(x,\nu)}$ is represented by $\rho_t^{(x,\nu)}/\rho_t^{(x,\nu)}(1)$, where

$$\rho_t^{(x,\nu)}(f) = \int E_Q\left[f(\varphi_t(y))\exp\left\{\sum_i \int_0^t h^i(\varphi_s(y))\,dY_s^i - \frac{1}{2}\int_0^t \sum_i h^i(\varphi_s(y))^2\,ds\right\}\right]d\nu(y)$$

and $Y_t = \int_0^t h\left(x_s^{(x)}\right) ds + N_t$. See [7]. Then $\rho_t^{(x_n,\nu_n)}(f)$ converges to $\rho_t^{(x,\nu)}(f)$ a.s. Therefore $\nu_t^{(x_n,\nu_n)}$ converges to $\nu_t^{(x,\nu)}$ a.s. The proof is complete.

## 2. Ergodic properties of the filtering process

We discuss the ergodic properties of the semigroup of the pair process $(x_t, \nu_t)$. The properties are concerned with the existence and the uniqueness of the invariant measures of the process $(x_t, \nu_t)$. We first discuss a property of an invariant measure, assuming its existence.

**Theorem 2.1.** *Suppose that the pair process $(x_t, \nu_t)$ has an invariant measure $\Phi$. Then:*

(i) *The marginal distribution $\Lambda(dx) = \Phi(dx, \mathcal{M}_1)$ is an invariant measure of the first component process $x_t$.*

(ii) *Suppose further that $\Phi$ satisfies*

$$\int \Phi(dx, d\nu)\, f(x)\, F(\nu) = \int \Phi\left(\mathbb{R}^d, d\nu\right) \nu(f)\, F(\nu) \tag{2.1}$$

*for any bounded $f$ and $F$. Then the marginal distribution $\tilde{\Phi}(d\nu) = \Phi\left(\mathbb{R}^d, d\nu\right)$ is an invariant measure of the second component process $\nu_t$.*

Before the proof, we remark that condition (2.1) is equivalent to condition (1.6) if $(x_0, \nu_0)$ is a random variable with joint law $\Phi$.

*Proof.* The first assertion (i) is obvious. To show (ii), let $(x_t, \nu_t)$ be the Markov process with transition probability $\Pi_t$ and initial distribution $\Phi(dx, d\nu)$. Then $\nu_t$ is a Markov process with transition probability $\tilde{\Pi}_t$ by Theorem 1.2, because condition (1.6) is satisfied. Therefore the marginal distribution $\tilde{\Phi}$ is an invariant measure of $\tilde{\Pi}_t$. The proof is complete.

We next show the existence of a $\Pi_t$-invariant probability assuming that the first component $x_t$ has an invariant probability $\Lambda$. We denote by $\bar{x}_t, t \in \mathbb{R}$ a stationary Markov process with the same transition probability as that of $x_t$ satisfying $P(\bar{x}_t \in A) = \Lambda(A)$ for any $t$ and $A$. Let $\bar{N}_t, t \in \mathbb{R}$ be a standard $m$-dimensional Brownian motion independent of $\bar{x}_t$. The stationary observation process $\bar{Y}_t$ is defined by

$$\bar{Y}_t - \bar{Y}_s = \int_s^t h(\bar{x}_u)\, du + \bar{N}_t - \bar{N}_s. \tag{2.2}$$

Set for $-\infty < s < t < \infty$

$$\mathcal{B}_{s,t} = \sigma(\bar{x}_u; s \le u \le t), \quad \mathcal{G}_{s,t} = \bigcap_{\varepsilon > 0} \sigma\left(\bar{Y}_u - \bar{Y}_v; s \le u, v \le t + \varepsilon\right) \tag{2.3}$$

and define two stationary filtering processes $\bar{\pi}_t^{(0)}$ and $\bar{\pi}_t^{(1)}$ by

$$\bar{\pi}_t^{(0)}(f) = E\left[f(\bar{x}_t) \,|\, \mathcal{G}_{-\infty,t}\right], \tag{2.4}$$

$$\bar{\pi}_t^{(1)}(f) = E\left[f(\bar{x}_t) \,|\, \mathcal{G}_{-\infty,t} \vee \mathcal{B}_{-\infty,-\infty}\right], \tag{2.5}$$

where $\mathcal{B}_{-\infty,-\infty} = \bigcap_{t \in \mathbb{R}} \mathcal{B}_{-\infty,t}$. Then both $\left(\bar{x}_t, \bar{\pi}_t^{(0)}\right)$ and $\left(\bar{x}_t, \bar{\pi}_t^{(1)}\right)$ are stationary processes with transition probabilities $\{\Pi_t\}$. Then the laws of $\left(\bar{x}_0, \bar{\pi}_0^{(0)}\right)$ and $\left(\bar{x}_0, \bar{\pi}_0^{(1)}\right)$ denoted by $\Phi^{(0)}$ and $\Phi^{(1)}$ respectively, are invariant measures. Further, since

$$E\left[f\left(\bar{x}_0^{(i)}\right) | \sigma\left(\bar{\pi}_0^{(i)}\right)\right] = \bar{\pi}_0^{(i)}(f)$$

holds a.s., both $\Phi^{(0)}$ and $\Phi^{(1)}$ satisfy property (2.1).

We wish to show that if $\Phi^{(0)} = \Phi^{(1)}$ is satisfied, it is the unique invariant measure with the marginal law $\Lambda$ satisfying (2.1).

Before we proceed, we introduce some notions related to convex functions. Let $F$ be a real function on $\mathbb{R}^d \times \mathcal{M}_1$. It is called convex with respect to $\nu$ if for every $x$, the inequality $F(x, \lambda\nu_1 + (1-\lambda)\nu_2) \le \lambda F(x,\nu_1) + (1-\lambda)F(x,\nu_2)$ is satisfied for any $0 \le \lambda < 1$ and $\nu_1, \nu_2$ of $\mathcal{M}_1$. Let $\Phi$ and $\Psi$ be two measures on $\mathbb{R}^d \times \mathcal{M}_1$. If $\Phi(F) \le \Psi(F)$ holds for any $F$ which is continuous, bounded from below and is convex with respect to $\nu$, it is denoted by $\Phi\alpha\Psi$. Then $\alpha$ is an order relation. Indeed, if $\Psi\alpha\Phi$ and $\Phi\alpha\Psi$ are satisfied, then $\Phi(F) = \Psi(F)$ holds for any $F$ which is written as the difference of two continuous functions bounded from below, convex with respect to $\nu$. Since the set of such functions is rich enough, $\Phi$ and $\Psi$ are identical. See Meyer [10].

**Theorem 2.2.** *Let $\Lambda$ be an invariant probability of the semigroup $\{P_t\}$.*

(i) *$\int \Pi_t(x, \Lambda, \cdot)\Lambda(dx)$ converges weakly to $\Phi^{(0)}$. Further it is minimal among $\Pi_t$-invariant measures with marginal law $\Lambda$ satisfying (2.1).*
(ii) *$\int \Pi_t(x, \delta_x, \cdot)\Lambda(dx)$ converges weakly to $\Phi^{(1)}$. Further it is maximal among $\Pi_t$-invariant measures with marginal law $\Lambda$ satisfying (2.1).*

For the proof of the theorem, we will prepare two lemmas. The first one is an extension of Jensen's inequality concerning conditional expectations.

**Lemma 2.3.** *Let $\pi$ be an $\mathcal{M}_1$-valued random variable and $\mathcal{G}$ be a sub $\sigma$-field. The conditional expectation of $\pi$ with respect to $\mathcal{G}$ denoted by $E[\pi|\mathcal{G}]$ is defined as an $\mathcal{M}_1$-valued $\mathcal{G}$-measurable random variable $\pi'$ such that $E[G(\pi)|\mathcal{G}] = G(\pi')$ holds for any continuous affine function $G$ on $\mathcal{M}_1$. Let $F(x,\nu)$ be a continuous function bounded from below and be convex with respect to $\nu$. Let $x_0$ be a $\mathcal{G}$-measurable $\mathbb{R}^d$-valued random variable. Then it holds*

$$F(x_0, E[\pi|\mathcal{G}]) \le E[F(x_0,\pi)|\mathcal{G}]. \tag{2.6}$$

The proof is immediate. See [5].

**Lemma 2.4.** *Let $F(x,\nu)$ be a continuous function bounded from below and be convex with respect to $\nu$. Let $\Phi$ be a probability on $\mathbb{R}^d \times \mathcal{M}_1$ satisfying (2.1). Set $\mu(A) = \Phi(A \times \mathcal{M}_1)$. Then*

$$\int \Pi_t F(x,\mu)\,\mu(dx) \le \int \Pi_t F(x,\nu)\,\Phi(dx,d\nu) \le \int \Pi_t F(x,\delta_x)\,\mu(dx). \tag{2.7}$$

*Proof.* Let $(x_0, \nu_0)$ be a random variable with the distribution $\Phi$. Let $\left(x_t^{(x_0)}, \nu_t^{(x_0,\nu_0)}\right)$ be the solution of equations (1.1) and (1.4). Then we have

$$E\left[\nu_t^{(x_0,\nu_0)}|\sigma(B_s, N_s; s \le t, x_0)\right] = \nu_t^{(x_0,\mu)}.$$

Therefore if $F$ is convex with respect to $\nu$, we have

$$F\left(x_t^{(x_0)}, \nu_t^{(x_0,\mu)}\right) \le E\left[F\left(x_t^{(x_0)}, \nu_t^{(x_0,\nu_0)}\right)|\sigma(B_s, N_s; s \le t, x_0)\right]$$

by Lemma 2.3. Taking expectations, we get the first inequality of (2.7).

Next let $y_0$ be an $\mathbf{R}^d$-valued random variable with the law $\mu$ independent of $B_t, N_t; t > 0,\ x_0$ satisfying $E[f(y_0)|\sigma(\nu_0)] = \nu_0(f)$. Then

$$E\left[\nu_t^{(x_0,\delta_{y_0})}|\sigma(B_s, N_s; s \le t, x_0, \nu_0)\right] = \nu_t^{(x_0,\nu_0)}$$

holds. Then we can prove the second inequality of (2.7) similarly as the above. The proof is complete.

*Proof of Theorem 2.2.* Let $\bar{x}_t, t \in \mathbf{R}$ be the stationary Markov process such that $P(\bar{x}_0 \in A) = \Lambda(A)$. Set $\bar{\Pi}_{-t,0}^{(0)}(f) = E\left[f(\bar{x}_0)|\mathcal{G}_{-t,0}\right]$, where $\mathcal{G}_{-t,0}$ is the $\sigma$-field defined by (2.3). Then we have

$$E\left[F\left(\bar{x}_0, \bar{\Pi}_{-t,0}^{(0)}\right)\right] = \int \pi_t F(x, \Lambda)\, \Lambda(dx).$$

Since $F\left(\bar{x}_0, \bar{\pi}_{-t,0}^{(0)}\right)$ converges to $F\left(\bar{x}_0, \bar{\pi}_0^{(0)}\right)$ a.s., the above converges to $\Phi^{(0)}(F)$. Next set $\bar{\pi}_{-t,0}^{(1)}(f) = E\left[f(\bar{x}_0)|\mathcal{G}_{-t,0} \vee \mathcal{B}_{-\infty,-t}\right]$. Then

$$E\left[F\left(\bar{x}_0, \bar{\pi}_{-t,0}^{(0)}\right)\right] = \int \Pi_t F(x, \delta_x)\, \Lambda(dx).$$

Since $F\left(\bar{x}_0, \bar{\pi}_{-t,0}^{(1)}\right)$ converges to $F\left(\bar{x}_0, \bar{\pi}_0^{(1)}\right)$ a.s., the above converges to $\Phi^{(1)}(F)$.

Finally let $\Phi$ be an arbitrary invariant probability with the marginal law $\Lambda$ satisfying (2.1). Let $F$ be a continuous function bounded from below and convex with respect to $\nu$. Then we have by Lemma 2.4.

$$\int \Pi_t F(x,\Lambda)\,\Lambda(dx) \le \int \Pi_t F(x,\nu)\,\Phi(dx,d\nu) \le \int \Pi_t F(x,\delta_x)\,\Lambda(dx).$$

Let $t$ tend to infinity. Then we obtain $\Phi^{(0)}(F) \le \Phi(F) \le \Phi^{(1)}(F)$. Therefore $\Phi^{(0)}$ is the minimal and $\Phi^{(1)}$ is the maximal one among $\Pi_t$-invariant measures with the marginal law $\Lambda$ satisfying (2.1). The proof is complete.

By Theorem 2.2, the uniqueness of $\Pi_t$-invariant measure with the marginal law $\Lambda$ satisfying (2.1) is justified if $\Phi^{(0)} = \Phi^{(1)}$ is satisfied. A sufficient condition for this is that the remote past $\sigma$-field $\mathcal{B}_{-\infty,-\infty}$ of the stationary process $\bar{x}_t$ is trivial, i.e., any $B$ of $\mathcal{B}_{-\infty,-\infty}$ fulfills $P(B) = 0$ or 1. The process $\bar{x}_t$ is then called *purely nondeterministic.* Now it is known that the process $\bar{x}_t$ is purely nondeterministic if and only if

$$\int |P_t f(x) - \Lambda(f)|\,d\Lambda(x) \underset{t\to\infty}{\longrightarrow} 0 \tag{2.8}$$

is satisfied for any $f$ of $L^1(\Lambda)$. Therefore we have the following corollary.

**Corollary 2.5.** *Assume (2.8) is satisfied for any $f$ of $L^1(\Lambda)$. Then the $\Pi_t$-invariant probability with marginal law $\Lambda$ satisfying (2.1) is unique and is equal to $\Phi^{(0)} = \Phi^{(1)}$.*

*Remark.* Assertions stronger than Theorem 2.2 and Corollary 2.5 are valid for $\tilde{\Pi}_t$-invariant measures. Let $\tilde{\Phi}^{(0)}$ and $\tilde{\Phi}^{(1)}$ be the laws of $\bar{\pi}_t^{(0)}$ and $\bar{\pi}_t^{(1)}$, respectively. Then $\tilde{\Phi}^{(0)}$ is the minimal one and $\tilde{\Phi}^{(1)}$ is the maximal one among all $\tilde{\Pi}_t$-invariant probabilities. Here a condition like (2.1) is not necessary. Indeed, instead of (2.7), we have

$$\tilde{\Pi}_t F(\mu) \le \int \tilde{\Pi}_t F(\nu)\tilde{\Phi}(d\nu) \le \int \tilde{\Pi}_t F(\delta_x)\mu(dx) \tag{2.9}$$

for any bounded continuous convex function $F$ on $\mathcal{M}_1$, where

$$\mu(f) = \int \Phi(d\nu)\nu(f).$$

See [5]. Therefore if the stationary process $\bar{x}_t$ is purely nondeterministic, then $\tilde{\Pi}_t$ has a unique invariant probability.

Finally we discuss a strong ergodic property of the semigroup $\Pi_t$. Assumptions required for this are slightly stronger than (2.8).

**Theorem 2.6.** *Assume that the system process has an invariant probability $\Lambda$ and that*

$$\exists \lim_{t\to\infty} P_t f(x) = \Lambda(f) \quad \forall x \in Supp(\Lambda) \tag{2.10}$$

*holds for any continuous function such that $\Lambda(|f|) < \infty$. Then the pair process $(x_t, \nu_t)$ has a unique invariant probability $\Phi$ with the marginal law $\Lambda$ satisfying (2.1). Further, if $Supp(\nu)$ is included in $Supp(\Lambda)$,*

$$\exists \lim_{t\uparrow\infty} \int \Pi_t F(x,\nu)\,\nu(dx) = \Phi(F) \tag{2.11}$$

*holds for any continuous function $F$ bounded from below and convex with respect to $\nu$. The limit is finite if $\int F(x,\delta_x)\Lambda(dx) < \infty$.*

*Proof.* For a fixed $s > 0$ and $\nu$ of $\mathcal{M}_1$, set $\Phi = \int \Pi_s(x,\nu,\cdot)\,\nu(dx)$. Then its marginal law $\Phi(A\times\mathcal{M}_1)$ coincides with $\nu P_s = \int P_s(x,\cdot)\nu(dx)$. Apply Lemma 2.4. Then we have

$$\int \Pi_t F(x,\nu P_s)\,\nu P_s(dx) \le \int \Pi_{t+s} F(x,\nu)\,\nu(dx) \le \int \Pi_t F(x,\delta_x)\,\nu P_s(dx)$$

for any continuous function $F$, bounded from below and convex with respect to $\nu$. Let $s$ tend to infinity. Since $\nu P_s$ converges to the invariant measure $\Lambda$ by our assumption, we have

$$\begin{aligned}\int \Pi_t F(x,\Lambda)\,\Lambda(dx) &\le \underline{\lim}_{s\to\infty} \int \Pi_s F(x,\nu)\,\nu(dx)\\ &\le \overline{\lim}_{s\to\infty} \int \Pi_s F(x,\nu)\,\nu(dx) \le \int \Pi_t F(x,\delta_x)\,\Lambda(dx)\end{aligned}$$

Next let $t$ tend to infinity. Then we get

$$\Phi^{(0)}(F) \leq \underline{\lim}_{s \to \infty} \int \Pi_s F(x,\nu)\nu(dx)$$
$$\leq \overline{\lim}_{s \to \infty} \int \Pi_s F(x,\nu)\nu(dx) \leq \Phi^{(1)}(F)$$

Since $\Phi^{(0)} = \Phi^{(1)}$ holds, we get (2.11). The limit is finite if $\int F(x, \delta_x)\, \Lambda(dx)$ is finite, since $\Phi^{(1)}(F)$ is bounded by the above quantity.

**Corollary 2.7.** *Assume the same condition as in Theorem 2.6. Let $x_t$ be a system process such that $x_0$ is in $Supp(\Lambda)$ a.s. Then*

$$\exists \lim_{t \to \infty} E\left[|f(x_t) - \pi_t(f)|^p\right] = \int \Phi(dx, d\nu)\, |f(x) - \nu(f)|^p \tag{2.12}$$

*holds for any continuous function $f$. Further the limit is finite if $\Lambda(|f|^p) < \infty$.*

If the infinitesimal generator $L$ of the system process is non-degenerate, a criterion that it has an invariant probability is known. See Hashiminsky [2]. Further in this case the property (2.10) is always satisfied. See [7].

We will consider the case that the operator $L$ is degenerate. For two vector fields $X, Y$, the Lie bracket $[X, Y] = XY - YX$ is again a vector field. The least $\mathbb{R}$-module including $X_0 + (\partial/\partial t), X_1, \ldots, X_n$ closed by the Lie bracket operations is denoted by $\mathcal{L}(X_0 + (\partial/\partial t), X_1, \ldots, X_n)$. Suppose now that the system process has an extremal invariant probability $\Lambda$. If the operator $L + (\partial/\partial t)$ satisfies Hörmander's hypoellipticity condition, namely, if

$$\dim \mathcal{L}\left(X_0 + \tfrac{\partial}{\partial t}, X_1, \ldots, X_n\right) = d + 1 \quad \forall x, t \tag{2.13}$$

is satisfied, then the transition probability $P_t(x, \cdot)$ has a smooth density and the convergence (2.10) holds on the support of the measure $\Lambda$. Therefore Corollary 2.7 is valid. However if (2.13) is not satisfied then the convergence (2.10) does not hold. The stationary system process is not purely nondeterministic. Therefore the filtering process may have two different $\tilde{\Pi}_t$-invariant measures satisfying (2.1). See Ichihara-Kunita [3].

**Theorem 2.8.** *Suppose that the system process satisfies (2.13) and has a unique invariant measure. Then the assertion of Corollary 2.7 is valid. Further the limit is positive if $f$ is not a constant on the support of $\Lambda$.*

*Proof.* The first assertion of the theorem is already shown above. In the sequel we prove the second assertion. Consider the stationary process $\left(\bar{x}_t, \bar{\pi}_t^{(0)}\right)$ introduced before. Since $\bar{x}_t$ satisfies the stochastic differential equation (1.1), we have

$$f(\bar{x}_t) = f(\bar{x}_0) + \int_0^t Lf(\bar{x}_s)\,ds + \sum_j \int_0^t X_j f(\bar{x}_s)\,dB_s^j. \tag{2.14}$$

Also from (1.5), $\bar{\pi}_t(f) \equiv \bar{\pi}_t^{(0)}(f)$ satisfies

$$\bar{\pi}_t(f) = \bar{\pi}_0(f) + \int_0^t \bar{\pi}_s(Lf)\,ds + \sum_i \int_0^t \left(\bar{\pi}_s\left(fh^i\right) - \bar{\pi}_s(f)\,\bar{\pi}_s\left(h^i\right)\right) d\bar{I}_s^i, \tag{2.15}$$

where $\bar{I}_t^i - \bar{I}_0^i \equiv \bar{Y}_t^i - \bar{Y}_0^i - \int_0^t \bar{\pi}_s\left(h^i\right) ds,\ i = 1, \dots, m$ is a standard Brownian motion independent of $B_t$. Now if (2.12) is 0, $f(x_t) = \bar{\pi}_t(f)$ is satisfied for any $t$. Then the terms involving the stochastic integrals in (2.14) and (2.15) must be 0. Therefore $X_j f(\bar{x}_0) = 0$ a.s. and hence $X_j f = 0$ on the support of $\Lambda$ for $j = 1, \dots, n$.

We claim $[X_j, X_0] f = 0,\ j = 1, \dots, n$. Since $X_0 X_j f = 0$, it is enough to prove $X_j X_0 f = 0,\ j = 1, \dots, n$. Since $Lf = X_0 f$, we have from (2.14),

$$f(\bar{x}_t) - f(\bar{x}_0) - \int_0^t X_0 f(\bar{x}_s)\,ds = 0 \qquad \text{a.s. } P. \tag{2.16}$$

Let $\varphi_t$ be the deterministic flow generated by $X_0$. Then we have

$$f(\varphi_t(\bar{x}_0)) - f(\bar{x}_0) - \int_0^t X_0 f(\varphi_s(\bar{x}_0))\,ds = 0 \qquad \text{a.s. } P. \tag{2.17}$$

The two equalities (2.16) and (2.17) imply $f(\bar{x}_t) = f(\varphi_t(\bar{x}_0))$ and $X_0 f(\bar{x}_s) = X_0 f(\varphi_s(\bar{x}_0))$ a.s. Consequently $X_0 f(\bar{x}_t)$ is a process of bounded variation. Since (2.14) is valid for $X_0 f(\bar{x}_t)$ instead of $f$, the martingale part

$\Sigma_j \int_0^t X_j X_0 f(\bar{x}_s)\, dB_s^j$ is 0, proving $X_j X_0 f = 0$ on the support of $\Lambda$, $j = 1, \ldots, n$. Now we can show by induction that $Zf = 0$ holds on the support of $\Lambda$ for any $Z$ of $\mathcal{L}(X_0 + (\partial/\partial t), X_1, \ldots, X_n)$. Then $f$ is a constant on the support of $\Lambda$ by the assumption (2.13). The proof is complete.

*Example.* (Linear filter) It is interesting to compare Theorem 2.8 with the Kalman filter. Suppose that the system process is defined by a linear stochastic differential equation

$$x_t = x_0 + GB_t + \int_0^t F x_s ds, \tag{2.18}$$

where $x_0$ is a Gaussian random variable independent of $B_t$ and $F, G$ are constant matrices. Then $x_t$ is a Gaussian Markov process. Its transition probability $P_t(x, \cdot)$ is a normal distribution $N(m_t, V_t)$ where the mean $m_t$ and the covariance $V_t$ are given by

$$m_t = e^{Ft} x, \quad V_t = \int_0^t e^{F(t-s)} G G' e^{F'(t-s)} ds. \tag{2.19}$$

The observation process is defined by

$$Y_t = \int_0^t H x_s ds + N_t, \tag{2.20}$$

where $H$ is a constant matrix. The filter $\pi_t$ is then a stochastic process with values in normal distributions, which we denote by $N(\hat{x}_t, P_t)$. The mean $\hat{x}_t$ is again a Gaussian Markov process governed by a stochastic differential equation

$$\hat{x}_t = \hat{x}_0 + \int_0^t (F - P_s H' H)\, \hat{x}_s ds + \int_0^t P_s H' dY_s \tag{2.21}$$

and $P_t$ is a deterministic matrix satisfying a Riccati differential equation

$$\frac{d}{dt}P_t = P_tF' + FP_t - P_tH'HP_t + GG'. \tag{2.22}$$

The matrix $P_t$ coincides with the error covariance $E[(x_t - \hat{x}_t)(x_t - \hat{x}_t)']$.

We will assume for a moment that the real parts of the characteristic values of the matrix $F$ are all negative. Then $m_t$ of (2.19) converges to 0 and $V_t$ converges to a finite matrix $V_\infty$. Therefore the transition probability $P_t(x, \cdot)$ of the system process converges weakly to a normal distribution $N(0, V_\infty)$ for any $x$, which is the unique invariant measure of the system process. Further the associated system process is purely nondeterministic. Then the pair process $(x_t, N(\hat{x}_t, P_t))$ has a unique invariant measure with the marginal law $N(0, V_\infty)$ satisfying (2.1). It is characterized as the law of a random variable $(x_\infty, N(\hat{x}_\infty, P_\infty))$, where $(x_\infty, \hat{x}_\infty)$ is a Gaussian random variable with values in $\mathbb{R}^{2d}$ with mean vector 0 and $Cov(x_\infty) = V_\infty, Cov(x_\infty, \hat{x}_\infty) = Cov(\hat{x}_\infty) = V_\infty - P_\infty$, where $P_\infty = \lim_{t\uparrow\infty} P_t$ (exists).

Now in the linear system (2.18) the vector fields $X_0, X_1, \cdots, X_n$ are given by

$$X_0 = \sum_{i,j} (f_{ij}x^j) \frac{\partial}{\partial x^i}, \quad X_j = \sum_i \sigma^i_j \frac{\partial}{\partial x^i}, \tag{2.23}$$

where $F = (f_{ij})$ and $G = (\sigma^i_j)$. A simple computation yields that (2.13) holds if and only if

$$rank(F, FG, \cdots, F^{d-1}G) = d. \tag{2.24}$$

In control theory it is known that the above holds if and only if the following control system is completely controllable

$$\frac{dx}{dt} = Fx + Gu(t), \tag{2.25}$$

i.e., for any two points $x_0, y_0$ of $\mathbb{R}^d$ and $t_0 > 0$, there exists a control $u(t)$ such that the solution starting at $x_0$ at time 0 reaches $y_0$ at time $t_0$. Furthermore under the condition (2.24), $\lim_{t\uparrow\infty} P_t = P_\infty$ exists and is strictly positive definite. (It could be infinite if some of the real parts of the characteristic values of F are nonnegative.) Assume further that the observation

$$Y_t = Hx_t \tag{2.26}$$

is completely observable, i.e., we can identify the state $x_t$ from the observation $Y_t$. Then $P_\infty$ is a finite matrix.

## 3. Upper and lower estimates of the filtering errors

In the previous section we saw that the error covariance of the linear filter converges to a finite and strictly positive definite matrix as the time tends to infinity provided that the associated control system is completely controllable and completely observable. For the nonlinear filter, a similar result is valid if the system process has an invariant probability as we saw it in Theorem 2.8. However if the system process does not have an invariant probability, such an ergodic property is not known. In this section, instead of trying to obtain the limiting error covariance, we shall obtain its upper bound and lower bound.

The system process, the observation process and the filtering process are the ones defined in Section 1. Let $f, g$ be continuous functions of polynomial growth. The error covariance of $f(x_t)$ and $g(x_t)$ is defined by

$$P_t(f,g) = E[(f(x_t) - \pi_t(f))(g(x_t) - \pi_t(g))]. \tag{3.1}$$

We first obtain a differential equation governing $P_t(f,g)$.

**Lemma 3.1.** *Assume that $f, g$ are $C^2$-functions such that their derivatives are of polynomial growth. Then $P_t(f,g)$ is differentiable with respect to $t$ and the derivative satisfies*

$$\begin{aligned} \tfrac{d}{dt} P_t(f,g) = P_t(f, Lg) + P_t(Lf,g) + Q_t(f,g) \\ - \sum_i E\big[\pi_t \left\{(f - \pi_t(f)) \left(h^i - \pi_t(h^i)\right)\right\} \\ \times \pi_t \left\{(g - \pi_t(g)) \left(h^i - \pi_t(h^i)\right)\right\}\big], \end{aligned} \tag{3.2}$$

*where*

$$Q_t(f,g) = E\left[\sum_{i,j} a_{ij}(x_t) \tfrac{\partial f}{\partial x^i}(x_t) \tfrac{\partial g}{\partial x^j}(x_t)\right]. \tag{3.3}$$

*Proof.* Note the equality

$$Lfg = fLg + gLf + \sum_{i,j} a_{ij} \tfrac{\partial f}{\partial x^i} \tfrac{\partial g}{\partial x^j}.$$

Then we have from the Kushner-Stratonovich equation

$$
\begin{aligned}
\pi_t(fg) = \pi_0(fg) &+ \int_0^t \pi_s(fLg)ds + \int_0^t \pi_s(gLf)ds \\
&+ \int_0^t \pi_s\left(\sum_{ij} a_{ij}\frac{\partial f}{\partial x^i}\frac{\partial g}{\partial x^j}\right)ds \\
&+ \mathcal{G}_t\text{-martingale with mean } 0.
\end{aligned}
\tag{3.4}
$$

Further, since $\pi_t(f)$ and $\pi_t(g)$ satisfy the Kushner-Stratonovich equation, we have by Itô's formula,

$$
\begin{aligned}
\pi_t(f)\pi_t(g) = \pi_0(f)\pi_0(g) &+ \int_0^t \pi_s(f)\pi_s(Lg)\,ds \\
&+ \int_0^t \pi_s(g)\pi_s(Lf)\,ds \\
&+ \sum_i \int_0^t \left\{\pi_s(fh^i) - \pi_s(f)\pi_s(h^i)\right\} \\
&\qquad\times \left\{\pi_s(gh^i) - \pi_s(g)\pi_s(h^i)\right\} ds \\
&+ \mathcal{G}_t\text{-martingale with mean } 0.
\end{aligned}
\tag{3.5}
$$

Note the relation $\pi_s(fg) - \pi_s(f)\pi_s(g) = \pi_s\{(f - \pi_s(f))(g - \pi_s(g))\}$. Then we have the equality $P_t(f,g) = E[\pi_t(fg) - \pi_t(f)\pi_t(g)]$. Therefore the two equalities (3.4) and (3.5) imply

$$
\begin{aligned}
P_t(f,g) = P_0(f,g) &+ \int_0^t \{P_s(f,Lg) + P_s(Lf,g) + Q_s(f,g)\}\,ds \\
&- \sum_i \int_0^t E\left[\pi_s\left\{(f - \pi_s(f))\left(h^i - \pi_s(h^i)\right)\right\}\right. \\
&\qquad\times \left.\pi_s\left\{(g - \pi_s(g))\left(h^i - \pi_s(h^i)\right)\right\}\right] ds.
\end{aligned}
$$

The proof is complete.

In the sequel we assume that the dimension of the system process and that of the observation process are the same and the map $h : \mathbf{R}^d \to \mathbf{R}^d$ is the identity, i.e. $h^i(x) = x^i$ holds for $i = 1, \ldots, d$. Set

$$(3.6) \qquad \begin{aligned} p_t &= E\left[|x_t - \hat{x}_t|^2\right], \quad q_t = \sum_i E\left[a^{ii}(x_t)\right], \\ a_t &= \sum_i E\left[\left(x_t^i - \hat{x}_t^i\right)\left(b^i(x_t) - \pi_t(b^i)\right)\right], \end{aligned}$$

$$(3.7) \qquad \tilde{p}_t = \sum_{i,k} E\left[\pi_t\left\{\left(x^i - \hat{x}_t^i\right)\left(x^k - \hat{x}_t^k\right)\right\}^2\right],$$

where $\hat{x}_t^i$ is the mean filter $\int x^i \pi_t(dx)$. Now observe that $Lx^i = b^i$, $i = 1, \ldots, n$ hold. Then Lemma 3.1 implies

$$(3.8) \qquad \frac{dp_t}{dt} = 2a_t + q_t - \tilde{p}_t.$$

The above is not a closed differential equation for $p_t$. We shall obtain a closed differential equation such that its solution dominates $p_t$. Observe first the inequality

$$(3.9) \qquad \tfrac{1}{d} p_t^2 \le \tilde{p}_t \le \hat{p}_t^2,$$

where

$$(3.10) \qquad \hat{p}_t = \mathrm{E}[|x_t - \hat{x}_t|^4]^{1/2}.$$

For $b^i, i = 1, \ldots, d$, we assume that there exists a real constant $c$ such that

$$(3.11) \qquad \sum_{i,j} \tfrac{\partial b^i}{\partial x^j}(x) \xi^i \xi^j \le c\,|\xi|^2$$

holds for any $x$ of $\mathbb{R}^d$ and real numbers $\xi^1, \dots, \xi^d$. Then using the mean value theorem, we have

$$\begin{aligned}(3.12)\qquad &\sum_i \pi_t \left\{ \left(b^i - \pi_t\left(b^i\right)\right)\left(x^i - \hat{x}_t^i\right)\right\} \\ &= \pi_t \left\{ \sum_{i,j} \frac{\partial b^i}{\partial x^j} \left(x^j - \hat{x}_t^j\right)\left(x^i - \hat{x}_t^i\right) \right\} \\ &\le \pi_t \left\{ c \sum_i \left|x^i - \hat{x}_t^i\right|^2 \right\}.\end{aligned}$$

Take the expectations of the above. Then we get $a_t \le c p_t$. Consequently we have the inequality

$$(3.13)\qquad \frac{dp_t}{dt} \le 2cp_t + q_t - \frac{1}{d} p_t^2.$$

By the comparison theorem of ordinary differential equations, $p_t$ is then dominated by the solution of the following differential equation,

$$(3.14)\qquad \frac{d\varphi_t}{dt} = 2c\varphi_t + q_t - \frac{1}{d}\varphi_t^2, \quad \varphi_0 = p_0.$$

We have thus the following upper estimate of $p_t$.

**Theorem 3.2.** *Assume (3.11). Let $\varphi_t$ be the solution of equation (3.14). Then $p_t \le \varphi_t$ is satisfied for any $t$.*

We next obtain the lower estimate for $\hat{p}_t$ defined by (3.10). We assume that there exists a constant $\hat{c}$ such that

$$(3.15)\qquad \sum_{i,j} \frac{\partial b^i}{\partial x^j}(x)\xi^i\xi^j \ge \hat{c}\,|\xi|^2$$

for any $x$ and real numbers $\xi^1, \dots, \xi^d$. Then we get $a_t \ge \hat{c} p_t$ similarly as before. Further if $\hat{c}$ is nonpositive we get $a_t \ge \hat{c}\hat{p}_t$. Therefore we have the inequality

$$(3.16)\qquad \hat{p} \ge p_0 + 2\int_0^t \hat{c}\hat{p}_s ds + \int_0^t q_s ds - \int_0^t \hat{p}_s^2 ds.$$

Then we can compare $\hat{p}_t$ with the solution of the following differential equation

$$\frac{d\hat{\varphi}_t}{dt} = 2\hat{c}\hat{\varphi}_t + q_t - \hat{\varphi}_t^2, \quad \hat{\varphi}_0 = p_0. \tag{3.17}$$

Indeed, set $\hat{\varphi}_t - \hat{p}_t = \psi_t$. It satisfies

$$\psi_t \leq 2\int_0^t \hat{c}\psi_s ds + \int_0^t q_s ds - \int_0^t (\hat{\varphi}_s + \hat{p}_s)\, \psi_s ds.$$

Then Gronwall's inequality implies $\psi_t \leq 0$. Therefore we have the following.

**Theorem 3.3.** *Assume (3.15) for some nonpositive constant $\hat{c}$. Let $\hat{\varphi}_t$ be the solution of equation (3.17). Then $\hat{p}_t \geq \hat{\varphi}_t$ is satisfied for any $t$.*

Finally we shall obtain the upper estimate of $\overline{\lim}_{t\to\infty} p_t$ and the lower estimate of $\underline{\lim}_{t\to\infty} \hat{p}_t$. We assume that the transition probabilities of the system process have strictly positive density functions.

Now the quantity $q_t$ defined by (3.6) is represented by $\nu(P_t Tr(a))$, where $Tr\,(a) = \sum_i a^{ii}$ and $\nu$ is the initial distribution of the system process. By the ergodic theorem for transition probabilities, we have the following.

(a) Suppose that $P_t$ has an invariant probability $\Lambda$. Then it holds

$$q_\infty = \exists \lim_{t\to\infty} q_t = \int_{\mathbb{R}^d} Tr(a)(x)\Lambda(dx).$$

(b) Suppose that $P_t$ does not have an invariant probability. If

$$\lim_{|x|\to\infty} Tr(a)(x)$$

exists, then it holds

$$q_\infty = \exists \lim_{t\to\infty} q_t = \lim_{|x|\to\infty} Tr(a)(x).$$

The proof is found in [7]. Then the nonnegative solution of equation (3.14) converges and the limit is a nonnegative solution of the following quadratic equation for $\alpha$:

$$2c\alpha + q_\infty - \tfrac{1}{d}\alpha^2 = 0. \tag{3.18}$$

Similarly the nonnegative solution of equation (3.17) converges and the limit is a nonnegative solution of the following quadratic equation for $\beta$:

$$2\hat{c}\beta + q_\infty - \beta^2 = 0. \tag{3.19}$$

From Theorems 3.2, 3.3 and the above argument, we have the following.

**Theorem 3.4.** *Suppose* $q_\infty = \lim_{t\uparrow\infty} q_t$ *exists.*

(i) *If* $0 < q_\infty < \infty$, *both equations (3.18) and (3.19) have unique positive solutions* $\alpha_0$ *and* $\beta_0$ *respectively. Further, it holds that* $\overline{\lim}_{t\to\infty} p_t \le \alpha_0$, *and* $\underline{\lim}_{t\to\infty} \hat{p}_t \ge \beta_0$.

(ii) *If* $q_\infty = 0$, *it holds that* $\overline{\lim}_{t\to\infty} p_t \le \max(0, 2cd)$. *In particular* $\overline{\lim}_{t\to\infty} p_t = 0$ *if* $c \le 0$.

(iii) *If* $q_\infty = \infty$, *it holds that* $\underline{\lim}_{t\to\infty} \hat{p}_t = \infty$.

## References

[1] M. Fujisaki, G. Kallianpur and H. Kunita, *Stochastic differential equations for the nonlinear filtering problem*, Osaka J. Math. **9** (1972), pp. 19–40.

[2] R.Z. Hashiminsky, *Ergodic properties of recurrent diffusion processes and stabilization of the solution of the Cauchy problem for parabolic equations*, Theor. Prob. Appl. **5** (1960), pp. 179–196.

[3] K. Ichihara and H. Kunita, *A classification of the second order degenerate elliptic operators and its probabilistic characterization*, Z. Wahrscheinlichkeitstheorie verw. Gebiete **30** (1974), pp. 235–254.

[4] R.E. Kalman and R.S. Bucy, *New results in linear filtering and prediction theory*, Trans. ASME Ser. D, J. Basic Eng. **83** (1961), pp. 95–108.

[5] H. Kunita, *Asymptotic behavior of the nonlinear filtering errors of Markov processes*, J. Multivariate Analysis **1** (1971), pp. 365–393.

[6] H. Kunita, *Stochastic partial differential equations connected with non-linear filtering*, "Nonlinear filtering and stochastic control," Lect. Notes Math., 1982, pp. 100–169.

[7] H. Kunita, *Stochastic Flows and Stochastic Differential Equations*, in preparation.

[8] T.G. Kurtz and D.L. Ocone, *Unique characterization of conditional distributions in nonlinear filtering*, The Annals of Probabilitiy **16** (1988), pp. 80–107.

[9] R. Sh. Liptser and A.N. Shiryaev, "Statistics of Random Processes I," Springer Verlag, New York, 1977.

[10] P.A. Meyer, "Probability and Potentials," Blaisdel, Waltham, Massachusetts, 1966.

Department of Applied Science
Kyushu University 36
Fukuoka, 812 Japan

## Progress in Probability



*Progress in Probability* is designed for the publication of workshops, seminars and conference proceedings on all aspects of probability theory and stochastic processes, as well as their connections with and applications to other areas such as mathematical statistics and statistical physics. It acts as a companion series to *Probability and Its Applications*, a context for research level monographs and advanced graduate texts.

We encourage preparation of manuscripts in some form of TeX for delivery in camera-ready copy, which leads to rapid publication, or in electronic form for interfacing with laser printers or typesetters.

Proposals should be sent directly to the editors or to:
Birkhäuser Boston, 675 Massachusetts Avenue, Cambridge, MA 02139, U.S.A.

1 ÇINLAR/CHUNG/GETOOR. Seminar on Stochastic Processes, 1981
2 KESTEN. Percolation Theory for Mathematicians
3 ASMUSSEN/HERING. Branching Processes
4 CHUNG/WILLIAMS. Introduction to Stochastic Integration
5 ÇINLAR/CHUNG/GETOOR. Seminar on Stochastic Processes, 1982
6 BLOOMFIELD/STEIGER. Least Absolute Deviation
7 ÇINLAR/CHUNG/GETOOR. Seminar on Stochastic Processes, 1983
8 BOUGEROL/LACROIX. Products of Random Matrices with Application to Schrödinger Operator
9 ÇINLAR/CHUNG/GETOOR. Seminar on Stochastic Processes, 1984
10 KIFER. Ergodic Theory of Random Transformations
11 EBERLEIN/TAQQU. Dependence in Probability and Statistics
12 ÇINLAR/CHUNG/GETOOR. Seminar on Stochastic Processes, 1985
13 ÇINLAR/CHUNG/GETOOR/GLOVER. Seminar on Stochastic Processes, 1986
14 DEVROYE. A Course in Density Estimation
15 ÇINLAR/CHUNG/GETOOR/GLOVER. Seminar on Stochastic Processes, 1987
16 KIFER. Random Perturbations of Dynamical Systems
17 ÇINLAR/CHUNG/GETOOR/GLOVER. Seminar on Stochastic Processes, 1988
18 ÇINLAR/CHUNG/GETOOR/FITZSIMMONS/WILLIAMS. Seminar on Stochastic Processes, 1989
19 ALEXANDER/WATKINS. Spatial Stochastic Processes: A Festschrift in Honor of Ted Harris on His 70th Birthday
20 HAAGERUP/HOFFMANN-JØRGENSEN/NIELSEN. Probability in Banach Spaces 6: Proceedings of the Sixth International Conference, Sandbjerg, Denmark 1986

21 Eberlein/Kuelbs/Marcus. Probability in Banach Spaces 7: Proceedings of the Seventh International Conference

22 Pinsky. Diffusion Processes and Related Problems in Analysis, Volume I: Diffusions in Analysis and Geometry

MIX
Papier aus verantwortungsvollen Quellen
Paper from responsible sources
FSC® C105338

If you have any concerns about our products,
you can contact us on
**ProductSafety@springernature.com**

In case Publisher is established outside the EU,
the EU authorized representative is:
**Springer Nature Customer Service Center GmbH**
**Europaplatz 3, 69115 Heidelberg, Germany**

Printed by Libri Plureos GmbH
in Hamburg, Germany